Essential Permutations & Combinations

A Self-Teaching Guide

Tim Hill

Questing Vole Press

Essential Permutations & Combinations: A Self-Teaching Guide
by Tim Hill

Editor: Kevin Debenjak
Proofreader: Diane Yee
Compositor: Kim Frees
Cover: Questing Vole Press

Contents

1 The Sum Rule and Product Rule

We start with two basic principles of counting: the sum rule and product rule. The statements of these rules are simple, as are the initial examples. More-complex problems can often be broken down into parts that can be solved by using these basic principles. With practice, you'll develop the ability to decompose such problems and then assemble the partial solutions to arrive at the final answer. With this goal in mind, our learn-by-example approach is to analyze and solve many diverse counting problems, noting the principles being used.

Our first principle of counting is the sum rule.

Sum Rule If a first task can be done in m ways, and a second task can be done in n ways, and the two tasks can't be performed simultaneously, then doing either task can be accomplished in any one of $m + n$ ways.

Throughout this book, when we say that a particular event, such as a first task, can arise in m ways, then these m ways are assumed to be distinct, unless otherwise stated.

Example 1.1 A library has 30 biology books and 40 anthropology books. By the sum rule a reader can select among $30 + 40 = 70$ books to learn more about one or the other of these two subjects.

Example 1.2   The sum rule can be generalized beyond two tasks provided that no pair of tasks can occur simultaneously.

A bookseller who has six different travel books each on Canada, Belgium, and France can recommend any one of these 18 books to a Francophone customer who's interested in traveling to one of these places.

Example 1.3   Professor Moriarty has three textbooks on criminology and Professor Plum has five such textbooks. If n denotes the maximum number of different books on this topic that a student can borrow from them, then $5 \leq n \leq 8$ because both professors *might* own copies of the same textbook(s).

Example 1.4   This example introduces our second principle of counting.

To decide whether to expand a plant, a manager assigns 12 employees to two committees. Committee A consists of five members and is to investigate possible favorable results from such an expansion. The other seven employees, committee B, will investigate possible unfavorable results. Should the manager decide to speak to only one committee member before making his decision, then by the sum rule there are $5 + 7 = 12$ employees that he can consult for input. To be more unbiased, however, he decides to speak with a member of committee A on Monday, and then with a member of committee B on Tuesday, before reaching a decision. By using the following principle, he can select two such employees to speak with in $5 \times 7 = 35$ ways.

> **Product Rule**   If a procedure can be broken down into first and second stages, and if there are m possible outcomes for the first stage and if, for each of these outcomes, there are n possible outcomes for the second stage, then the total procedure can be carried out, in the designated order, in mn ways.

Example 1.5   A drama club is holding auditions for a play. If six men and eight women audition for the leading male and female roles, then by the product rule the director can cast the leading couple in $6 \times 8 = 48$ ways.

Example 1.6 This example illustrates extensions of the product rule.

A license plate has two letters followed by four digits.

(a) If no letter or digit can be repeated, then $26 \times 25 \times 10 \times 9 \times 8 \times 7 = 3{,}276{,}000$ different license plates are possible. Each of the 26 letters can occupy the first position. Because repetitions aren't allowed here, we can select only one of the 25 remaining letters to fill the second position. The same argument applies to the positions of the digits.

(b) If repetitions of letters and digits are allowed, then $26 \times 26 \times 10 \times 10 \times 10 \times 10 = 6{,}760{,}000$ different plates are possible.

(c) If repetitions are allowed, as in part (b), then how many of the plates have only vowels (A, E, I, O, U) and even digits (0, 2, 4, 6, 8)? See Problem 2 in Chapter 2.

Example 1.7 A computer's main memory contains a large collection of circuits, each of which is capable of storing a *bit*—that is, one of the *binary digits* 0 or 1. These storage circuits are arranged in units called *cells* (or memory cells). To identify the cells in memory, each is assigned a unique *address*.

(a) If a computer represents an address by an ordered list of eight bits, collectively referred to as a *byte*, then by the product rule there are $2 \times 2 \times 2 \times 2 \times 2 \times 2 \times 2 \times 2 = 2^8 = 256$ such bytes. So a programmer can use 256 addresses for cells to store information.

(b) If a computer uses two-byte addresses to identify memory cells, then such addresses are made up of two consecutive bytes, or 16 consecutive bits. Thus there are $256 \times 256 = 2^8 \times 2^8 = 2^{16} = 65{,}536$ available addresses that can be used to identify the cells in main memory.

(c) If a computer uses a four-byte (32-bit) addressing system, then there are $2^8 \times 2^8 \times 2^8 \times 2^8 = 2^{32} = 4{,}294{,}967{,}296$ available addresses that can be used to identify the cells in main memory.

(d) For an eight-byte (64-bit) architecture, each address comprises $8 \times 8 = 64$ bits, so $2^{64} = 18{,}446{,}744{,}073{,}709{,}551{,}616$ addresses are possible.

Example 1.8 At times it's necessary to combine different counting principles to solve a problem. In this example, both the sum and product rules are needed to get the answer.

A small coffee stand has a limited menu: six types of muffins, eight types of sandwiches, and five types of beverages (hot coffee, hot tea, iced tea, cola, and orange juice). For lunch, a customer wants either a muffin and a hot beverage or a sandwich and a cold beverage.

By the product rule there are $6 \times 2 = 12$ ways to purchase a muffin and hot beverage. Again by the product rule there are $8 \times 3 = 24$ possibilities for a sandwich and cold beverage. So by the sum rule there are $12 + 24 = 36$ ways in which the customer can buy lunch.

---

2   Permutations

Continuing to examine applications of the product rule, we turn to counting linear arrangements of objects. These arrangements are called **permutations** when the objects are distinct. We'll develop some systematic methods for counting linear arrangements, starting with a common example.

Example 2.1   In a class of 10 students, five are to be chosen and seated in a row for a photo. How many such linear arrangements are possible?

Solution   The key word here is *arrangement*, which states the importance of *order*. If A, B, C, ..., I, J denote the 10 students, then BCEFI, CEFIB, and ABCFG are three such different arrangements, even though the first two involve the same five students.

To answer this question, consider the positions and possible numbers of students that we can choose from to fill each position. The filling of a position is a stage of our procedure.

$$10 \quad \times \quad 9 \quad \times \quad 8 \quad \times \quad 7 \quad \times \quad 6$$

| 1st position | 2nd position | 3rd position | 4th position | 5th position |

Any of the 10 students can occupy the 1st position in the row. Because repetitions aren't possible here, we can select only one of the nine remaining students to fill the 2nd position. Continuing in this way, we find only six students to select from to fill the 5th and final position. This yields a total of 30,240 possible arrangements of five students selected from the class of 10.

Exactly the same answer is obtained if the positions are filled from right to left—namely, $6 \times 7 \times 8 \times 9 \times 10$. If the 3rd position is filled first, the 1st position second, the 4th position third, the 5th position fourth, and the 2nd position fifth, then the answer is $9 \times 6 \times 10 \times 8 \times 7$, still the same value, 30,240.

As illustrated in the preceding example, the product of consecutive positive integers is often needed to solve counting problems. The following notation lets us answer such problems concisely.

Definition For an integer $n \geq 0$, n **factorial** (denoted by $n!$) is defined by

$$0! = 1,$$
$$n! = (n)(n-1)(n-2) \cdots (3)(2)(1), \qquad \text{for } n \geq 1.$$

So $1! = 1$, $2! = 2$, $3! = 6$, $4! = 24$, and $5! = 120$. Also, for every $n \geq 0$, $(n+1)! = (n+1)(n!)$. We have sensible reasons for defining $0! = 1$, as you'll soon see.

Note that $n!$ grows rapidly. $10! = 3,628,800$ is exactly the number of seconds in six weeks, $11!$ exceeds the number of seconds in one year, $12!$ exceeds the number in 12 years, and $13!$ surpasses the number of seconds in a century.

By using factorial notation, the answer to Example 2.1 can be given more compactly as:

$$10 \times 9 \times 8 \times 7 \times 6 = 10 \times 9 \times 8 \times 7 \times 6 \times \frac{5 \times 4 \times 3 \times 2 \times 1}{5 \times 4 \times 3 \times 2 \times 1} = \frac{10!}{5!}.$$

Definition Given a collection of n distinct objects, any (linear) arrangement of these objects is called a **permutation** of the collection.

Given the letters a, b, c, we have six ways to arrange, or permute, all the letters: abc, acb, bac, bca, cab, cba. If we arrange only two of the letters at a time, then there are six such size-2 permutations: ab, ba, ac, ca, bc, cb.

If there are n distinct objects and r is an integer, where $1 \le r \le n$, then by the product rule the number of permutations of size r for the n objects is

$$P(n,r) \;=\; \underset{\substack{\text{1st}\\\text{position}}}{n} \;\times\; \underset{\substack{\text{2nd}\\\text{position}}}{(n-1)} \;\times\; \underset{\substack{\text{3rd}\\\text{position}}}{(n-2)} \;\times\cdots\times\; \underset{\substack{\text{rth}\\\text{position}}}{(n-r+1)}$$

$$= (n)(n-1)(n-2)\cdots(n-r+1)\times\frac{(n-r)(n-r-1)\cdots(3)(2)(1)}{(n-r)(n-r-1)\cdots(3)(2)(1)}$$

$$= \frac{n!}{(n-r)!}$$

A specific case of this result is Example 2.1, where $n = 10$, $r = 5$, and $P(10, 5) = 30{,}240$.

Note the following facts for $n!$:

- If $r = 0$, then $P(n, 0) = 1 = n!/(n - 0)!$, so $P(n, r) = n!/(n - r)!$ holds for all $0 \le r \le n$.

- If all n objects in a collection are permuted, then $r = n$, so $P(n, n) = n!/0! = n!$.

- If $n \ge 2$, then $P(n, 2) = n!/(n - 2)! = n(n - 1)$.

The number of permutations of size r, where $0 \le r \le n$, from a collection of n objects, is $P(n, r) = n!/(n - r)!$. Remember that $P(n, r)$ counts (linear) arrangements in which the objects *can't* be repeated. If repetitions are allowed, then by the product rule there are n^r possible arrangements, where $r \ge 0$.

Example 2.2 The number of permutations of the letters in the word BACKDROP is 8!. If only five of the letters are used, then the number of permutations (of size 5) is $P(8, 5) = 8!/(8 - 5)! = 8!/3! = 6720$. If repetitions of letters are allowed, then the number of possible 12-letter sequences is $8^{12} \approx 6.872 \times 10^{10}$.

Example 2.3 Unlike the preceding example, the number of (linear) arrangements of the four letters in BALL is 12, not 4! (= 24), because we don't have four *distinct* letters to arrange. The 12 arrangements are:

A B L L	L A B L
A L B L	L A L B
A L L B	L B A L
B A L L	L B L A
B L A L	L L A B
B L L A	L L B A

If the two L's are distinguished as L_1, L_2, then we can use our previous ideas on permutations of distinct objects; with the four distinct symbols B, A, L_1, L_2, we have 4! = 24 permutations:

A B L_1 L_2	A B L_2 L_1
A L_1 B L_2	A L_2 B L_1
A L_1 L_2 B	A L_2 L_1 B
B A L_1 L_2	B A L_2 L_1
B L_1 A L_2	B L_2 A L_1
B L_1 L_2 A	B L_2 L_1 A
L_1 A B L_2	L_2 A B L_1
L_1 A L_2 B	L_2 A L_1 B
L_1 B A L_2	L_2 B A L_1
L_1 B L_2 A	L_2 B L_1 A
L_1 L_2 A B	L_2 L_1 A B
L_1 L_2 B A	L_2 L_1 B A

The preceding lists show that for each arrangement in which the L's are indistinguishable there corresponds a pair of permutations with distinct L's. Consequently,

2 × (Number of arrangements of the letters B, A, L, L) = (Number of permutations of the symbols B, A, L_1, L_2),

so the answer to the original problem of finding all the arrangements of the four letters in BALL is 4!/2 = 12.

Example 2.4 By using the idea developed in the preceding example, we can count the arrangements of the nine letters in DATABASES.

There are 3! = 6 arrangements with the A's distinguished for *each* arrangement in which the A's aren't distinguished. For example, $DA_1TA_2BA_3SES$, $DA_1TA_3BA_2SES$, $DA_2TA_1BA_3SES$, $DA_2TA_3BA_1SES$, $DA_3TA_1BA_2SES$, and $DA_3TA_2BA_1SES$ all correspond to DATABASES when we remove the subscripts of the A's. Also, to the arrangement $DA_1TA_2BA_3SES$ there corresponds the pair of permutations $DA_1TA_2BA_3S_1ES_2$ and $DA_1TA_2BA_3S_2ES_1$ when the S's are distinguished. Consequently,

(2!)(3!)(Number of arrangements of the letters in DATABASES) = (Number of permutations of the symbols D, A_1, T, A_2, B, A_3, S_1, E, S_2),

so the number of arrangements of the nine letters in DATABASES is $9!/(2!3!) = 30{,}240$.

Before stating a general principle for arrangements with repeated symbols, note that in the preceding two examples we solved a new type of problem by invoking previous counting principles. This practice is common in mathematics in general, and often occurs in the derivations of counting formulas.

If there are n objects with n_1 indistinguishable objects of a first type, n_2 indistinguishable objects of a second type,..., and n_r indistinguishable objects of an rth type, where $n_1 + n_2 + \cdots + n_r = n$, then there are

$$\frac{n!}{n_1! n_2! \cdots n_r!}$$ (linear) arrangements of the given n objects.

Example 2.5 Arranging all the letters in MASSASAUGA, there are $\dfrac{10!}{4!\,3!\,1!\,1!\,1!} = 25{,}200$ possible arrangements. Among these are

$\dfrac{7!}{3!\,1!\,1!\,1!\,1!} = 840$ in which all four A's are together. To get the preceding result, we counted all arrangements of the seven symbols AAAA (one symbol), S, S, S, M, U, G.

Example 2.6 Determine the number of *staircase* paths in the xy-plane from (2, 1) to (7, 4), where each such path is a series of individual steps going one unit rightward (R) or one unit upward (U). The heavy lines in the following figure show two of these paths.

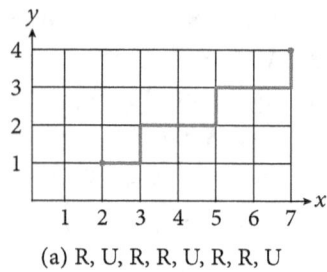

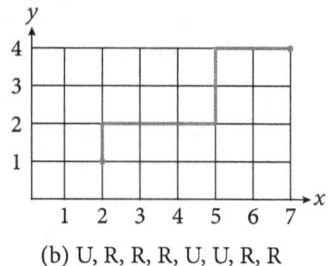

(a) R, U, R, R, U, R, R, U   (b) U, R, R, R, U, U, R, R

Below each path in the figure is a list of individual steps. For example, in part (a) the list R, U, R, R, U, R, R, U indicates that starting at the point (2, 1), we first move one unit rightward [to (3, 1)], then one unit upward [to (3, 2)], followed by two units rightward [to (5, 2)], and so on, until we reach the point (7, 4). The path consists of five R's for moves rightward and three U's for moves upward.

The path in part (b) of the figure is also made up of five R's and three U's. In general, the overall trip from (2, 1) to (7, 4) requires $7 - 2 = 5$ horizontal moves rightward and $4 - 1 = 3$ vertical moves upward. Consequently, each path corresponds to a list of five R's and three U's, and the solution for the number of paths is the number of arrangements of the five R's and three U's, which is $8!/(5!3!) = 56$.

Example 2.7 We now become a bit more abstract and prove that if n and k are positive integers where $n = 2k$, then $n!/2^k$ is an integer. Because our argument relies on counting, it's an example of a **combinatorial proof**.

Consider then symbols $x_1, x_1, x_2, x_2, \ldots, x_k, x_k$. The number of ways in which we can arrange all these $n = 2k$ symbols is an integer that equals

$$\underbrace{\frac{n!}{2!\,2!\cdots 2!}}_{k \text{ factors of } 2!} = \frac{n!}{2^k}.$$

Example 2.8 In this example, we apply our ideas so far to a situation in which the arrangements aren't linear.

If six people, designated as A, B,..., F, are seated about a round table, then how many different circular arrangements are possible, if arrangements are considered to be the same when one can be obtained from the other by rotation?

Solution In the following figure, arrangements (a) and (b) are considered to be identical (that is, (a) can be rotated to arrive at (b)), whereas (b), (c), and (d) are three distinct arrangements.

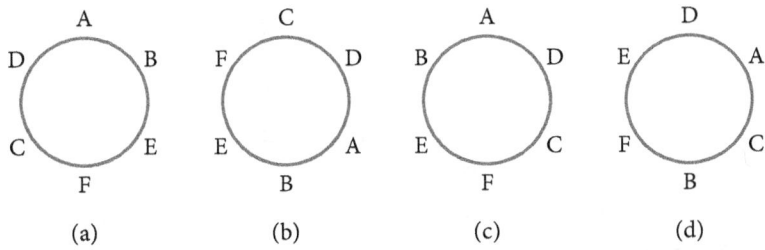

Let's relate this problem to earlier ones. Consider parts (a) and (b) in the figure. Starting at the top of the circle and moving clockwise, we list the distinct linear arrangements ABEFCD and CDABEF, which correspond to the same circular arrangement. In addition to these two, four other linear arrangements—BEFCDA, DABEFC, EFCDAB, and FCDABE—correspond to the same circular arrangement as (a) or (b). So because each circular arrangement corresponds to six linear arrangements, we have 6 × (Number of circular arrangements of A, B,..., F) = (Number of linear arrangements of A, B,..., F) = 6!.

Consequently, there are 6!/6 = 5! = 120 arrangements of A, B,..., F around a circular table.

Example 2.9 Suppose that the six people in Example 2.8 are three women (A, B, and C) and three men (D, E, and F). In how many ways can we arrange these six people around the table so that the sexes alternate? (Again, arrangements are considered to be identical if one can be obtained from the other via rotation.)

Solution   Before we solve this problem, let's solve Example 2.8 by using an alternative method, which will help in solving the present problem. If we place A at the table as shown in part (a) of the following figure, then five positions (clockwise from A) remain to be filled. Using B, C,..., F to fill these five positions is the problem of permuting B, C,..., F in a linear manner, which can be done in 5! = 120 ways.

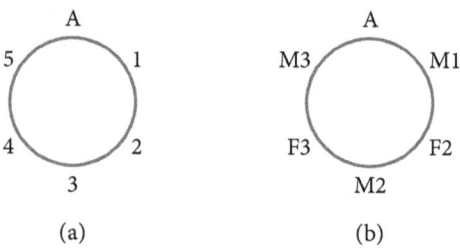

(a)   (b)

To solve the new problem of alternating the sexes, consider the method shown in part (b) of the figure. A (a female) is placed as before. The next position, clockwise from A, is marked M1 (Male 1) and can be filled in three ways. Continuing clockwise from A, position F2 (Female 2) can be filled in two ways. Proceeding in this manner, by the product rule there are $3 \times 2 \times 2 \times 1 \times 1 = 12$ ways in which these six people can be arranged with no two men or women seated next to each other.

Stirling's Formula

For large values of n, it can be computationally expensive or nearly impossible to compute $n!$ exactly. For $n \geq 70$, $n! > 10^{100}$ and can't be represented on many scientific calculators. In most cases for which $n!$ is needed with a large value of n, we need only the ratio of $n!$ to another large number a_n. A common example of this is $P(n, r)$ with large n and not so large r, which equals $n!/(n - r)!$. In such cases, we can observe that

$$\frac{n!}{a_n} = e^{\log(n!) - \log(a_n)}.$$

Compared to computing $n!$, it takes a much larger n before $\log(n!)$ becomes difficult to represent. Furthermore, if we had a simple approximation s_n to $\log(n!)$ such that

$$\lim_{n \to \infty} |s_n - \log(n!)| = 0,$$

then the ratio of $n!/a_n$ to s_n/a_n would be close to 1 for large n. The following result, whose proof can be found in William Feller, *An Introduction to Probability Theory and Its Applications, Vol. 1*, provides such an approximation.

Stirling's Formula   Let

$$s_n = \frac{1}{2}\log(2\pi) + \left(n + \frac{1}{2}\right)\log(n) - n.$$

Then

$$\lim_{n\to\infty}\left|s_n - \log(n!)\right| = 0.$$

Put another way

$$\lim_{n\to\infty}\frac{(2\pi)^{1/2}n^{n+1/2}e^{-n}}{n!} = 1.$$

Example 2.10   Approximate $P(70, 20) = 70!/50!$ by using Stirling's formula.

Solution   The approximation from Stirling's formula is

$$\frac{70!}{50!} \approx \frac{(2\pi)^{1/2}70^{70.5}e^{-70}}{(2\pi)^{1/2}50^{50.5}e^{-50}} = 3.940\times10^{35}.$$

The exact calculation yields 3.938×10^{35}. The approximation and the exact calculation differ by less than 1/10 of 1 percent.

Problems

1. Eight candidates from party A and five candidates from party B are nominated for president of a school board.

    (a) If the president is to be one of these candidates, then how many possibilities are there for the winner?

    (b) How many possibilities exist for a pair of candidates (one from each party) to oppose each other for the election?

    (c) Which counting principle is used in part (a)? in part (b)?

2. Answer part (c) of Example 1.6 in Chapter 1.

3. A brand of automobile comes in four models, 12 colors, three engine sizes, and two transmission types.

   (a) How many distinct automobiles can be manufactured?

   (b) If one of the available colors is blue, then how many different blue automobiles can be manufactured?

4. The board of directors of a construction company has 10 members. An upcoming stockholders' meeting is scheduled to approve a new slate of company officers chosen from the 10 board members.

   (a) How many different slates consisting of a president, vice president, secretary, and treasurer can the board present to the stockholders for their approval?

   (b) Three members of the board of directors are engineers. How many slates from part (a) have (i) an engineer nominated for the presidency? (ii) exactly one engineer appearing on the slate? (iii) at least one engineer appearing on the slate?

5. Betty and Veronica witness a car speeding away from the front of a jewelry store as a burglar alarm sounds. They give the police the following information about the car's license plate, which consists of two letters followed by four digits. Betty says that the second letter on the plate is either an O or a Q and the last digit is either a 3 or an 8. Veronica says that the first letter on the plate is either a C or a G and that the first digit is a 7. How many different license plates must the police investigate?

6. A race awards different-sized trophies to the first eight finishers.

   (a) If 30 people enter the race, then in how many ways is it possible to award the trophies?

   (b) If Chris and David are two participants in the race, then in how many ways can the trophies be awarded with these two runners among the top three?

7. A restaurant advertises that a customer can buy a hamburger with or without any or all of the following: ketchup, mustard, mayonnaise,

lettuce, tomato, onion, pickle, cheese, or mushrooms. How many different kinds of hamburger orders are possible?

8. A database administrator is managing a queue of 12 programs that have been submitted for batch processing. In how many ways can he order the processing of these programs if:

(a) there are no restrictions?

(b) he considers four of the programs to be higher in priority than the other eight and wants to process those four first?

(c) he first separates the programs into four of top priority, five of lesser priority, and three of least priority, and he wants to process the 12 programs in such a way that the top-priority programs are processed first and the three least-priority programs are processed last?

9. A bakery offers eight different kinds of pastry and six different kinds of muffins. In addition to bakery items, customers can purchase small, medium, or large containers of the following beverages: coffee (black, with cream, with sugar, or with cream and sugar), tea (plain, with cream, with sugar, with cream and sugar, with lemon, or with lemon and sugar), hot cocoa, and orange juice. In how many ways can a customer order:

(a) one bakery item and one medium-sized beverage for herself?

(b) one bakery item and one container of coffee for herself and one muffin and one container of tea for her son?

(c) one pastry and one container of tea for herself, one muffin and one container of orange juice for her son, and one bakery item and one container of coffee for each of her parents?

10. Alice has 15 different books. In how many ways can she place her books on two shelves so that there is at least one book on each shelf? (Consider the books in each arrangement to be placed one next to the other, with the first book on each shelf at the left of the shelf.)

11. Three small towns, designated by A, B, and C, are connected by a system of two-way roads, as shown in the following figure.

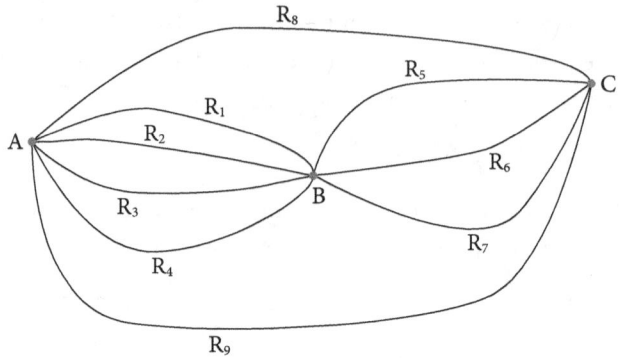

(a) In how many ways can you travel from town A to town C?

(b) How many different round trips can you travel from town A to town C and back to town A?

(c) How many of the round trips in part (b) are such that the return trip (from town C to town A) differs at least partially from the route you take from town A to town C? (For example, if you travel from town A to town C along roads R_1 and R_6, then on your return you might take roads R_6 and R_3 , or roads R_7 and R_2, or road R_9, among other possibilities, but you *don't* travel on roads R_6 *and* R_1.)

12. List all the permutations for the letters a, c, t.

13. (a) How many permutations are there for the eight letters a, c, f, g, i, t, w, x?

(b) Consider the permutations in part (a). How many start with the letter t? How many start with the letter t and end with the letter c?

14. Evaluate:
    (a) $P(7, 2)$
    (b) $P(8, 4)$
    (c) $P(10, 7)$
    (d) $P(12, 3)$

15. In how many ways can the symbols a, b, c, d, e, e, e, e, e be arranged so that no e is adjacent to another e?

16. An alphabet of 40 symbols is used to transmit messages in a communication system.

    (a) How many distinct messages (lists of symbols) of 25 symbols can the transmitter generate if symbols can be repeated in the message?

    (b) How many if 10 of the 40 symbols can appear only as the first or last (or both) symbols of the message, the other 30 symbols can appear anywhere, and repetitions of all symbols are allowed?

17. On the internet each network interface of a computer is assigned one or more internet addresses. The type of these internet addresses depends on network size. According to the internet standard for reserved network numbers (STD 2), each address is a 32-bit string of one of the following three classes:

    (1) A class A address, meant for large networks, starts with the one-bit string 0, followed by a seven-bit *network number*, and then a 24-bit *local address*. Network numbers of all 0's or all 1's and local addresses of all 0's or all 1's are prohibited.

    (2) A class B address, meant for mid-sized networks, starts with the two-bit string 10, followed by a 14-bit network number, and then a 16-bit local address. Local addresses of all 0's or all 1's are prohibited.

    (3) A class C address, meant for small networks, starts with the three-bit string 110, followed by a 21-bit network number, and then an eight-bit local address. Local addresses of all 0's or all 1's are prohibited.

    For this internet standard, how many different addresses of each class are available on the internet?

18. Daisy is going to buy of a laptop computer. Her research indicates that seven models meet her needs. Furthermore, she'll also buy one of four mice, one of three external hard drives, and one of six

printers. (Here each peripheral device of a given type, such as the printer, is compatible with all seven laptops.) In how many ways can Daisy configure her computer system?

19. A programmer has seven different programming books on a bookshelf. Three of the books are about C; the other four are about Python. In how many ways can the programmer arrange these books on the shelf if:

(a) there are no restrictions?

(b) the languages should alternate?

(c) all the C books must be next to each other?

(d) all the C books must be next to each other and all the Python books must be next to each other?

20. (a) In how many ways can the letters in DATABASE be arranged?

(b) In how many of the arrangements in part (a) are all three A's together?

21. (a) In how many ways can the letters in SOCIOLOGICAL be arranged?

(b) In how many of the arrangements in part (a) are A and G adjacent?

(c) In how many of the arrangements in part (a) are all the vowels adjacent?

22. How many positive integers n can be formed by using the digits 3, 4, 4, 5, 5, 6, 7 if n must exceed 5,000,000?

23. Twelve clay targets (identical in shape) are arranged in four hanging columns, as shown in the following figure. Four red targets are in the first column, three white targets are in the second column, two green targets are in the third column, and three blue targets are in the fourth column.

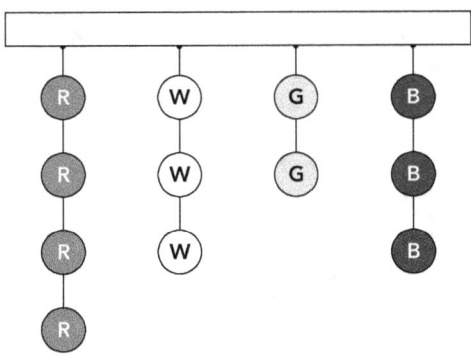

A competitive shooter must break all 12 targets (using a 12-round pistol), always breaking the existing target at the bottom of a column. Under these conditions, in how many different orders can the shooter shoot (and break) the 12 targets?

24. Show that for all integers $n, r \geq 0$, if $n + 1 > r$, then

$$P(n+1,r) = \left(\frac{n+1}{n+1-r} \right) P(n,r).$$

25. Find the value(s) of n in:
    (a) $P(n, 2) = 90$
    (b) $P(n, 3) = 3 \cdot P(n, 2)$
    (c) $2 \cdot P(n, 2) + 50 = P(2n, 2)$

26. How many different paths in the xy-plane are there from $(0, 0)$ to $(7, 7)$ if a path proceeds one step at a time by going either one space rightward (R) or one space upward (U)? How many such paths are there from $(2, 7)$ to $(9, 14)$? Make a general statement about these two results.

27. (a) How many distinct paths are there from $(-1, 2, 0)$ to $(1, 3, 7)$ in Euclidean three-space if each move is one of the following types?
    (H): $(x, y, z) \rightarrow (x + 1, y, z)$;
    (V): $(x, y, z) \rightarrow (x, y + 1, z)$;
    (A): $(x, y, z) \rightarrow (x, y, z + 1)$

    (b) How many such paths are there from $(1, 0, 5)$ to $(8, 1, 7)$?

    (c) Generalize the results in parts (a) and (b).

28. (a) Determine the value of the integer variable *counter* after execution of the following program snippet. (Here i, j, and k are integer variables.)

```
counter := 0
for i := 1 to 12 do
   counter := counter + 1
for j := 5 to 10 do
   counter := counter + 2
for k := 15 downto 8 do
   counter := counter + 3
```

(b) Which counting principle is used in part (a)?

29. Consider the following program snippet where i, j, and k are integer variables.

```
for i := 1 to 12 do
   for j := 5 to 10 do
      for k := 15 downto 8 do
         print (i - j) * k
```

(a) How many times is the print statement executed?

(b) Which counting principle is used in part (a)?

30. A sequence of letters of the form abcba, where the expression is unchanged upon reversing order, is an example of a *palindrome* (of five letters).

(a) If a letter can appear more than twice, then how many palindromes of five letters are there? of six letters?

(b) Repeat part (a) under the condition that no letter appears more than twice.

31. (a) Determine the number of six-digit integers (no leading zeros) in which no digit can be repeated.

(b) Answer part (a) if digits can be repeated.

(c) Answer parts (a) and (b) with the extra condition that the six-digit integer is (i) even; (ii) divisible by 5; (iii) divisible by 4 (Hint: An integer is divisible by 4 if and only if the integer formed by the last two digits is divisible by 4).

32. (a) Provide a combinatorial argument to show that if n and k are positive integers where $n = 3k$, then $n!/(3!)^k$ is an integer.

(b) Generalize the result of part (a).

33. (a) In how many ways can a student answer all the questions on a 10-question true–false test?

(b) In how many ways can the student answer the test in part (a) if it's possible to leave a question unanswered to avoid an extra penalty for a wrong answer?

34. How many distinct four-digit integers can be made from the digits 1, 3, 3, 7, 7, and 8?

35. (a) In how many ways can seven people be arranged about a circular table?

(b) If two of the people insist on sitting next to each other, then how many arrangements are possible?

36. (a) In how many ways can eight people, denoted by A, B,..., H, be seated about the square table shown in the following figure, where tables (a) and (b) are considered to be the same seating but are distinct from table (c)?

(b) How many different seatings are possible with A and B not sitting next to each other?

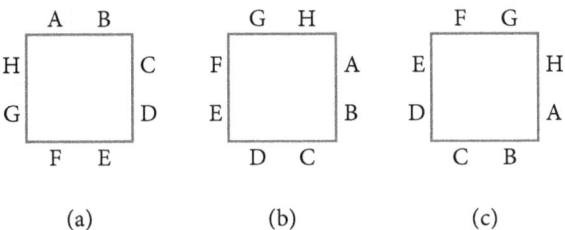

(a)   (b)   (c)

3

Combinations

The standard deck of playing cards consists of 52 cards comprising four suits: clubs (♣), diamonds (♦), hearts (♥), and spades (♠). Each suit has 13 cards: ace, 2, 3,..., 9, 10, jack, queen, king. If we draw three cards from a standard deck, in succession and without replacement, then by the product rule there are

$$52 \times 51 \times 50 = \frac{52!}{49!} = P(52, 3)$$

possibilities, one of which is A♥ (ace of hearts), 9♣ (nine of clubs), K♦ (king of diamonds). If instead we simply select three cards at one time from the deck so that the order of selection of the cards is no longer important, then the six permutations A♥9♣K♦, A♥K♦9♣, 9♣A♥K♦, 9♣K♦A♥, K♦9♣A♥, and K♦A♥9♣ all correspond to only one (unordered) selection. Consequently, each selection, or combination, of three cards, *with no regard to order*, corresponds to 3! permutations of three cards. In equation form this translates into

$$(3!) \times (\text{Number of selections of size 3 from a deck of 52})$$

$$= \text{Number of permutations of size 3 for the 52 cards}$$

$$= P(52, 3) = \frac{52!}{49!}.$$

Consequently, three cards can be drawn, without replacement, from a standard deck in 52!/(3!49!) = 22,100 ways.

If we start with n distinct objects, then each **selection**, or **combination**, of r of these objects, with no reference to order, corresponds to $r!$ permutations of size r from the n objects. Thus the number of combinations of size r from a collection of size n is

$$C(n, r) = \frac{P(n, r)}{r!} = \frac{n!}{r!(n-r)!}, \quad 0 \le r \le n.$$

In addition to $C(n, r)$ the symbol $\binom{n}{r}$ is also frequently used. Both $C(n, r)$ and $\binom{n}{r}$ are read "n choose r". Note that for all $n \ge 0$, $C(n, 0) = C(n, n) = 1$. Also, for all $n \ge 1$, $C(n, 1) = C(n, n-1) = n$. If $0 \le n < r$, then $C(n, r) = \binom{n}{r} = 0$.

When solving any counting problem, always determine the importance of order in the problem. When order is relevant, think in terms of permutations [$P(n, r)$] and arrangements (n^r) and the product rule (Chapter 1). When order isn't relevant, combinations typically play a key role in solving the problem.

Example 3.1 A hostess is having a dinner party for some members of her social committee. She has a small home and can invite only 11 of the 20 committee members. Order is not important, so she can invite the 11 people in $C(20, 11) = \binom{20}{11} = 20!/(11!9!) = 167{,}960$ ways.

Example 3.2 Alice and Betty decide to buy a lottery ticket together. To win the grand prize one must match five numbers selected from 1 to 49 inclusive and then also must match a sixth number, an integer selected from 1 to 42 inclusive. Alice selects the five numbers (between 1 and 49 inclusive), which she can do in $\binom{49}{5}$ ways (because matching doesn't involve order). Meanwhile Betty selects the sixth number, which she can do in $\binom{42}{1}$ ways. So by the product rule Alice and Betty can select the six numbers for their lottery ticket in $\binom{49}{5}\binom{42}{1} = 80{,}089{,}128$ ways.

Example 3.3 (a) A student taking a test must answer any seven of 10 essay questions. Order is unimportant, so the student can complete the test in

$$\binom{10}{7} = \frac{10!}{7!3!} = \frac{10 \times 9 \times 8}{3 \times 2 \times 1} = 120 \text{ ways.}$$

(b) If the student must answer three questions from the first five and four questions from the last five, then three questions can be selected from the first five in $\binom{5}{3} = 10$ ways, and the other four questions can be selected in $\binom{5}{4} = 5$ ways. So by the product rule the student can complete the test in $\binom{5}{3}\binom{5}{4} = 10 \times 5 = 50$ ways.

(c) If the student must answer seven of the 10 questions where at least three are selected from the first five, then we have three cases to consider:

(i) The student answers three of the first five questions and four of the last five questions. By the product rule this case can happen in $\binom{5}{3}\binom{5}{4} = 10 \times 5 = 50$ ways, as in part (b).

(ii) The student answers four of the first five questions and three of the last five questions. By the product rule this case can happen in $\binom{5}{4}\binom{5}{3} = 5 \times 10 = 50$ ways.

(iii) The student answers all five of the first five questions and two of the last five questions. By the product rule this case can happen in $\binom{5}{5}\binom{5}{2} = 1 \times 10 = 10$ ways.

Combining the results for cases (i), (ii), and (iii), by the sum rule the student can make $\binom{5}{3}\binom{5}{4} + \binom{5}{4}\binom{5}{3} + \binom{5}{5}\binom{5}{2} = 50 + 50 + 10 = 110$ selections of seven (out of 10) questions where each selection includes at least three of the first five questions.

Example 3.4 (a) A high-school gym teacher must select nine girls from the junior and senior classes for a volleyball team. If there are 28 juniors and 25 seniors, then the teacher can select the team in $\binom{53}{9} = 4{,}431{,}613{,}550$ ways.

(b) If two juniors and one senior are the best spikers and must be on the team, then the rest of the team can be chosen in $\binom{50}{6} = 15{,}890{,}700$ ways.

(c) For a certain tournament the team must comprise four juniors and five seniors. The teacher can select the four juniors in $\binom{28}{4}$ ways. For each of these selections she has $\binom{25}{5}$ ways to choose the five seniors. So by the product rule she can select the team in $\binom{28}{4}\binom{25}{5} = 1{,}087{,}836{,}750$ ways for this tournament.

Example 3.5 Depending on how they're analyzed, certain problems can be solved by using either arrangements or combinations, as illustrated by this example.

A gym teacher must make up four volleyball teams of nine girls each from the 36 freshman girls in her physical education class. In how many ways can she select these four teams? Call the teams A, B, C, and D.

(a) *Solution 1 (combinations)* To form team A, the teacher can select any nine girls from the 36 enrolled in $\binom{36}{9}$ ways. For team B she can select from the remaining 27 girls, yielding $\binom{27}{9}$ possibilities. This leaves $\binom{18}{9}$ and $\binom{9}{9}$ possible ways to select teams C and D, respectively. So by the product rule the four teams can be chosen in

$$\binom{36}{9}\binom{27}{9}\binom{18}{9}\binom{9}{9} = \left(\frac{36!}{9!\,27!}\right)\left(\frac{27!}{9!\,18!}\right)\left(\frac{18!}{9!\,9!}\right)\left(\frac{9!}{9!\,0!}\right)$$

$$= \frac{36!}{9!\,9!\,9!\,9!} \approx 2.145 \times 10^{19} \text{ ways.}$$

(b) *Solution 2 (arrangements)* For an alternative solution, consider the 36 students lined up as:

student 1, student 2, student 3,..., student 36

To select the four teams, distribute nine A's, nine B's, nine C's, and nine D's in the 36 slots. The number of ways in which this can be done is the number of arrangements of 36 letters comprising nine each of A, B, C, and D. This is now the familiar problem of arrangements of nondistinct objects, and the answer is

$$\frac{36!}{9!\,9!\,9!\,9!},$$

as in part (a).

Example 3.6 This example illustrates how some problems require the concepts of both arrangements and combinations for their solutions.

The number of arrangements of the letters in TALLAHASSEE is

$$\frac{11!}{3!\,2!\,2!\,2!\,1!\,1!} = 831,600.$$

How many of these arrangements have no adjacent A's?

Solution When we exclude the A's, there are

$$\frac{8!}{2!\,2!\,2!\,1!\,1!} = 5040$$

ways to arrange the remaining letters. One of these 5040 ways is shown in the following figure, in which the arrows indicate nine possible positions for the three A's.

Three of these positions can be selected in $\binom{9}{3} = 84$ ways, and because this is also possible for all the other 5039 arrangements of E, E, S, T, L, L, S, H, then by the product rule there are $5040 \times 84 = 423,360$ arrangements of the letters in TALLAHASSEE with no consecutive A's.

Before proceeding we need to introduce a concise way of writing the sum of a list of $n + 1$ terms like $a_m, a_{m+1}, a_{m+2}, \ldots, a_{m+n}$, where m and n are integers and $n \geq 0$. This notation is called **sigma notation** because it involves the capital Greek letter Σ (sigma, for *sum*); we use it to represent a summation by writing

$$a_m + a_{m+1} + a_{m+2} + \cdots + a_{m+n} = \sum_{i=m}^{m+n} a_i.$$

Here, the letter i is called the **index** of the summation, and this index spans all integers starting with the **lower limit** m and continuing up to and including the **upper limit** $m + n$.

We can use sigma notation as shown in the following examples.

$$\sum_{i=3}^{7} a_i = a_3 + a_4 + a_5 + a_6 + a_7 = \sum_{j=3}^{7} a_j.$$

The preceding example shows that there's nothing special about the letter i.

$$\sum_{i=1}^{4} i^2 = 1^2 + 2^2 + 3^2 + 4^2 = 30 = \sum_{k=0}^{4} k^2, \text{ because } 0^2 = 0.$$

$$\sum_{i=11}^{100} i^3 = 11^3 + 12^3 + 13^3 + \cdots + 100^3$$

$$= \sum_{j=12}^{101} (j-1)^3$$

$$= \sum_{k=10}^{99} (k+1)^3.$$

$$\sum_{i=7}^{10} 2i = 2(7) + 2(8) + 2(9) + 2(10) = 68$$

$$= 2(34) = 2(7 + 8 + 9 + 10)$$

$$= 2\sum_{i=7}^{10} i.$$

$$\sum_{i=3}^{3} a_i = a_3 = \sum_{i=4}^{4} a_{i-1} = \sum_{i=2}^{2} a_{i+1}.$$

$$\sum_{i=1}^{5} a = a + a + a + a + a = 5a.$$

By using this summation notation, we can express the answer to part (c) of Example 3.3 as

$$\binom{5}{3}\binom{5}{4} + \binom{5}{4}\binom{5}{3} + \binom{5}{5}\binom{5}{2} = \sum_{i=3}^{5}\binom{5}{i}\binom{5}{7-i}$$

$$= \sum_{j=2}^{4}\binom{5}{7-j}\binom{5}{j}.$$

Sigma notation will appear in many subsequent examples, including the next one.

Example 3.7   In computer science, certain arrangements, called *strings*, are made up from a prescribed *alphabet* of symbols. If the prescribed alphabet consists of the symbols 0, 1, and 2, for example, then 01, 11, 21, 12, and 20 are five of the 9 (3^2) strings of *length* 2. Among the 27 (3^3) strings of length 3 are 000, 012, 202, and 110.

In general, if n is any positive integer, then by the product rule there are 3^n strings of length n for the alphabet 0, 1, and 2. If $x = x_1 x_2 x_3 \cdots x_n$ is one of these strings, then we define the *weight* of x, denoted by wt(x), to be wt(x) = $x_1 + x_2 + x_3 + \cdots + x_n$. For example, wt(12) = 3 and wt(22) = 4 for $n = 2$; wt(101) = 2, wt(210) = 3, and wt(222) = 6 for $n = 3$.

Among the 3^{10} strings of length 10 for the alphabet 0, 1, and 2, how many have even weight?

Solution   Because the alphabet consists of only 0, 1, and 2, a string is even only if it contains exactly an even number of 1's. We have six different cases to consider. If the string x contains no 1's, then each of the 10 positions in x can be filled with either 0 or 2, and by the product rule there are 2^{10} such strings. If the string contains two 1's, then the positions for these two 1's can be selected in $\binom{10}{2}$ ways. After these two positions have been specified, there are 2^8 ways to place either 0 or 2 in the other eight positions. Hence there are $\binom{10}{2} \cdot 2^8$ strings of even weight that contain two 1's. The numbers of strings for the other four cases are given in the following table.

Number of 1's	Number of Strings
4	$\binom{10}{4}2^6$
6	$\binom{10}{6}2^4$
8	$\binom{10}{8}2^2$
10	$\binom{10}{10}2^0$

So by the sum rule the number of strings of length 10 that have even weight is

$$2^{10} + \binom{10}{2}2^8 + \binom{10}{4}2^6 + \binom{10}{6}2^4 + \binom{10}{8}2^2 + \binom{10}{10} = \sum_{n=0}^{5}\binom{10}{2n}2^{10-2n}.$$

Example 3.8 Beware of **overcounting**—a situation that often arises in easy-seeming counting problems when like selections are treated incorrectly as distinct. This example illustrates a way in which overcounting can lead us astray.

(a) Suppose that Alice draws five cards from a standard deck of 52 cards. In how many ways can she select a hand with no clubs? Here we want to count all five-card selections such as

A♥, 3♠, 4♠, 6♦, J♦,

5♠, 7♠, 10♠, 7♦, K♦, and

2♦, 3♦, 6♦, 10♦, J♦.

Alice is restricted to selecting her five cards from the 39 cards in the deck that aren't clubs, so she can make her selection in $\binom{39}{5}$ ways.

(b) Now suppose that we want to count the number of Alice's five-card selections that contain at least one club. These hands are precisely the selections that were *not* counted in part (a). And because there are $\binom{52}{5}$ possible five-card hands in total, then

$$\binom{52}{5} - \binom{39}{5} = 2{,}598{,}960 - 575{,}757 = 2{,}023{,}203$$

of all five-card hands contain at least one club.

(c) Let's try to solve part (b) in another way. Because Alice wants to have at least one club in the five-card hand, let her first select a club, which she can do in $\binom{13}{1}$ ways. Now she doesn't care what comes up for the other four cards. So after she eliminates the one club chosen from her standard deck, she can then select the other four cards in $\binom{51}{4}$ ways. So by the product rule the number of selections is

$$\binom{13}{1}\binom{51}{4} = 13 \times 249{,}900 = 3{,}248{,}700.$$

Something is *wrong* here. This answer is larger than that in part (b) by more than one million hands. Let's determine whether the mistake is in part (b) or in our current reasoning.

Suppose that Alice first selects 3♣ and then selects

5♣, K♣, 7♥, J♠.

If instead she first selects 5♣ and then selects

3♣, K♣, 7♥, J♠,

then does this selection actually differ from the prior selection? Clearly not. And the case where she first selects K♣ and then selects

3♣, 5♣, 7♥, J♠

doesn't differ from the preceding two selections.

Consequently, our current approach is wrong because we are overcounting; that is, erroneously treating like selections as if they were distinct.

(d) Fortunately, there's another way to arrive at the correct answer to part (b). Because the five-card hands must each contain at least one club, we have five cases to consider, listed in the following table. The table shows, for example, that $\binom{13}{2}\binom{39}{3}$ five-card hands contain exactly two clubs. If we want exactly three clubs in the hand, then the table indicates that $\binom{13}{3}\binom{39}{2}$ such hands are possible.

Number of Clubs	Number of Ways to Select This Number of Clubs	Number of Cards That Are Not Clubs	Number of Ways to Select This Number of Nonclubs
1	$\binom{13}{1}$	4	$\binom{39}{4}$
2	$\binom{13}{2}$	3	$\binom{39}{3}$
3	$\binom{13}{3}$	2	$\binom{39}{2}$
4	$\binom{13}{4}$	1	$\binom{39}{1}$
5	$\binom{13}{5}$	0	$\binom{39}{0}$

Because no two of the cases in the table have any five-card hand in common (hence no overcounting), the number of hands that Alice can select with at least one club is

$$\binom{13}{1}\binom{39}{4}+\binom{13}{2}\binom{39}{3}+\binom{13}{3}\binom{39}{2}+\binom{13}{4}\binom{39}{1}+\binom{13}{5}\binom{39}{0}$$

$$=\sum_{i=1}^{5}\binom{13}{i}\binom{39}{5-i}$$

$$=(13)(82{,}251) + (78)(9139) + (286)(741) +$$

$$(715)(39) + (1287)(1)$$

$$=2{,}023{,}203.$$

Problems

1. Calculate $\binom{6}{2}$ and check your answer by listing all the selections of size 2 that can be made from the letters a, b, c, d, e, and f.

2. Jane decides to take five magazines on a flight from the 12 that she has at home. In how many ways can she make her selection?

3. Evaluate:
   (a) $C(10, 4)$
   (b) $\binom{12}{7}$
   (c) $C(14, 12)$
   (d) $\binom{15}{10}$

4. In the Braille system a symbol—such as a lowercase letter or punctuation mark—is given by raising at least one of the dots in the six-dot (3 × 2) arrangement ⠿.
   (a) How many different symbols can the Braille system represent?
   (b) How many symbols have exactly three raised dots?
   (c) How many symbols have an even number of raised dots?

5. (a) How many *permutations* of size 3 can one produce with the letters m, r, a, f, and t?

   (b) List all the *combinations* of size 3 for the letters m, r, a, f, and t.

6. If n is a positive integer and $n > 1$, then prove that $\binom{n}{2} + \binom{n-1}{2}$ is a perfect square.

7. A committee of 12 is to be selected from 10 men and 10 women. In how many ways can the selection be made if:
   (a) there are no restrictions?
   (b) there must be six men and six women?
   (c) there must be an even number of women?
   (d) there must be more women than men?
   (e) there must be at least eight men?

8. In how many ways can a gambler draw five cards from a standard deck and get:
   (a) a flush (five cards of the same suit)?
   (b) four aces?
   (c) four of a kind?
   (d) three aces and two jacks?
   (e) three aces and a pair?
   (f) a full house (three of a kind and a pair)?
   (g) three of a kind?
   (h) two pairs?

9. How many bytes contain:
   (a) exactly two 1's?
   (b) exactly four 1's?
   (c) exactly six 1's?
   (d) at least six 1's?

10. (a) How many ways are there to pick a five-person basketball team from 12 possible players?

    (b) How many selections include the weakest player and the strongest player?

11. A student must answer seven out of 10 questions on an test. In how many ways can he make his selection if:
    (a) there are no restrictions?
    (b) he must answer the first two questions?
    (c) he must answer at least four of the first six questions?

12. In how many ways can 12 different books be distributed among four children so that:

    (a) each child gets three books?

    (b) the two oldest children get four books each and the two youngest get two books each?

13. How many arrangements of the letters in MISSISSIPPI have no consecutive S's?

14. A coach must select 11 seniors to play on a football team. If he can make his selection in 12,376 ways, then how many seniors are eligible to play?

15. (a) Fifteen points, no three of which are collinear, are given on a plane. How many lines do they determine?

    (b) Twenty-five points, no four of which are coplanar, are given in space. How many triangles do they determine? How many planes? How many tetrahedra (pyramid-like solids with four triangular faces)?

16. Determine the value of each of the following summations.

    (a) $\displaystyle\sum_{i=1}^{6}(i^2+1)$

    (b) $\displaystyle\sum_{j=-2}^{2}(j^3-1)$

    (c) $\displaystyle\sum_{i=0}^{10}[1+(-1)^i]$

    (d) $\displaystyle\sum_{k=n}^{2n}(-1)^k$, where n is an odd positive integer

    (e) $\displaystyle\sum_{i=1}^{6}i(-1)^i$

17. Express each of the following using the summation (sigma) notation. In parts (a), (d), and (e), n denotes a positive integer.

    (a) $\dfrac{1}{2!}+\dfrac{1}{3!}+\dfrac{1}{4!}+\cdots+\dfrac{1}{n!}, \quad n\geq 2$

    (b) $1+4+9+16+25+36+49$

    (c) $1^3-2^3+3^3-4^3+5^3-6^3+7^3$

    (d) $\dfrac{1}{n}+\dfrac{2}{n+1}+\dfrac{3}{n+2}+\cdots+\dfrac{n+1}{2n}$

    (e) $n-\left(\dfrac{n+1}{2!}\right)+\left(\dfrac{n+2}{4!}\right)-\left(\dfrac{n+3}{6!}\right)+\cdots+(-1)^n\left(\dfrac{2n}{(2n)!}\right)$

18. For the strings of length 10 in Example 3.7, how many have:
    (a) four 0's, three 1's, and three 2's?
    (b) at least eight 1's?
    (c) weight 4?

19. Consider the collection of all strings of length 10 made up from the alphabet 0, 1, 2, and 3.
    (a) How many of these strings have weight 3?
    (b) How many have weight 4?
    (c) How many have even weight?

4

The Binomial Theorem

This chapter examines three important results related to the concept of combinations.

First, for integers n, r, where $n \geq r \geq 0$, $\binom{n}{r} = \binom{n}{n-r}$. This fact can be proved algebraically by using the formula for $\binom{n}{r}$, but let's explain it intuitively: observe that when selecting r objects from a collection of n distinct objects, the selection process leaves behind $n - r$ objects. Consequently, $\binom{n}{r} = \binom{n}{n-r}$ affirms that a correspondence exists between the selections of size r (objects chosen) and the selections of size $n - r$ (objects left behind). An example of this correspondence is shown in the following table, where $n = 5$, $r = 2$, and the distinct objects are 1, 2, 3, 4, and 5.

Selection of Size $r = 2$ (Objects Chosen)		Selection of Size $n-r = 3$ (Objects Left Behind)	
1. 1,2	6. 2,4	1. 3,4,5	6. 1,3,5
2. 1,3	7. 2,5	2. 2,4,5	7. 1,3,4
3. 1,4	8. 3,4	3. 2,3,5	8. 1,2,5
4. 1,5	9. 3,5	4. 2,3,4	9. 1,2,4
5. 2,3	10. 4,5	5. 1,4,5	10. 1,2,3

You might be familiar with our second result from your past experience in algebra.

The Binomial Theorem If x and y are variables and n is a positive integer, then

$$(x+y)^n = \binom{n}{0}x^0 y^n + \binom{n}{1}x^1 y^{n-1} + \binom{n}{2}x^2 y^{n-2} + \cdots$$

$$+ \binom{n}{n-1}x^{n-1}y^1 + \binom{n}{n}x^n y^0 = \sum_{k=0}^{n}\binom{n}{k}x^k y^{n-k}.$$

Before we prove the general case, let's look at a special case. If $n = 4$, then the coefficient of $x^2 y^2$ in the expansion of the product

$$(x+y)(x+y)(x+y)(x+y)$$

1st   2nd   3rd   4th
factor   factor   factor   factor

is the number of ways in which we can select two x's from the four x's, one of which is available in each factor. Although the x's are identical in appearance, we distinguish them as the x in the first factor, the x in the second factor,..., and the x in the fourth factor. Also, note that when we select two x's, we use two factors, leaving us with two other factors from which we can select the two y's that are needed. For example, among the possibilities, we can select

x from the first two factors and y from the last two factors

or

x from the first and third factors and y from the second and fourth factors.

The following table summarizes the six possible selections. Consequently, the coefficient of $x^2 y^2$ in the expansion of $(x + y)^4$ is $\binom{4}{2} = 6$, the number of ways to select two distinct objects from a collection of four distinct objects.

Factors Selected for x		Factors Selected for y	
(1)	1,2	(1)	3,4
(2)	1,3	(2)	2,4
(3)	1,4	(3)	2,3
(4)	2,3	(4)	1,4
(5)	2,4	(5)	1,3
(6)	3,4	(6)	1,2

Now we prove the general case.

Proof In the expansion of the product

$$(x+y)(x+y)(x+y) \cdots (x+y)$$

1st factor 2nd factor 3rd factor nth factor

the coefficient of $x^k y^{n-k}$, where $0 \leq k \leq n$, is the number of different ways in which we can select k x's [and consequently $(n-k)$ y's] from the n available factors. (One way, for example, is to choose x from the first k factors and y from the last $n-k$ factors.) The total number of such selections of size k from a collection of size n is $C(n, k) = \binom{n}{k}$, and from this the binomial theorem follows. ∎

In view of this theorem, $\binom{n}{k}$ is often called a **binomial coefficient**.

Notice that it's also possible to express the result of the binomial theorem as

$$(x+y)^n = \sum_{k=0}^{n} \binom{n}{n-k} x^k y^{n-k}.$$

Example 4.1 (a) By the binomial theorem the coefficient of $x^5 y^2$ in the expansion of $(x + y)^7$ is $\binom{7}{5} = \binom{7}{2} = 21$.

(b) To get the coefficient of $a^5 b^2$ in the expansion of $(2a - 3b)^7$, replace $2a$ by x and $-3b$ by y. By the binomial theorem the coefficient of $x^5 y^2$ in $(x + y)^7$ is $\binom{7}{5}$, and $\binom{7}{5} x^5 y^2 = \binom{7}{5} (2a)^5 (-3b)^2 = \binom{7}{5} (2)^5 (-3)^2 a^5 b^2 = 6048 a^5 b^2$.

Corollary 4.1 For each integer $n > 0$,

(a) $\binom{n}{0} + \binom{n}{1} + \binom{n}{2} + \cdots + \binom{n}{n} = 2^n$, and

(b) $\binom{n}{0} - \binom{n}{1} + \binom{n}{2} - \cdots + (-1)^n \binom{n}{n} = 0$.

Proof Part (a) results from the binomial theorem when $x = y = 1$. Part (b) results when $x = -1$ and $y = 1$. ∎

Our third and final result, the multinomial theorem, generalizes the binomial theorem.

The Multinomial Theorem For positive integers n, t, the coefficient of $x_1^{n_1} x_2^{n_2} x_3^{n_3} \cdots x_t^{n_t}$ in the expansion of $(x_1 + x_2 + x_3 + \cdots + x_t)^n$ is

$$\frac{n!}{n_1! \, n_2! \, n_3! \cdots n_t!},$$

where each n_i is an integer with $0 \le n_i \le n$, for all $1 \le i \le t$, and $n_1 + n_2 + n_3 + \cdots + n_t = n$.

Proof As in the proof of the binomial theorem, the coefficient of $x_1^{n_1} x_2^{n_2} x_3^{n_3} \cdots x_t^{n_t}$ is the number of ways that we can select x_1 from n_1 of the n factors, x_2 from n_2 of the remaining $n - n_1$ factors, x_3 from n_3 of the now remaining $n - n_1 - n_2$ factors,..., and x_t from n_t of the final remaining $n - n_1 - n_2 - n_3 - \cdots - n_{t-1} = n_t$ factors. This can be carried out, as in part (a) of Example 3.5, in

$$\binom{n}{n_1}\binom{n-n_1}{n_2}\binom{n-n_1-n_2}{n_3}\cdots\binom{n-n_1-n_2-n_3-\cdots-n_{t-1}}{n_t}$$

ways. A bit of algebra (see Problem 3) shows that this product is equal to

$$\frac{n!}{n_1!\,n_2!\,n_3!\cdots n_t!},$$

which is also written as

$$\binom{n}{n_1, n_2, n_3, \ldots, n_t}$$

and is called a **multinomial coefficient**. (When $t = 2$ this expression reduces to a binomial coefficient.) ∎

Example 4.2   (a) By the multinomial theorem the coefficient of $x^2y^2z^3$ in the expansion of $(x + y + z)^7$ is $\binom{7}{2,2,3} = \frac{7!}{2!\,2!\,3!} = 210$, the coefficient of xyz^5 is $\binom{7}{1,1,5} = 42$, and the coefficient of x^3z^4 is $\binom{7}{3,0,4} = \frac{7!}{3!\,0!\,4!} = 35$.

(b) To get the coefficient of $a^2b^3c^2d^5$ in the expansion of $(a + 2b - 3c + 2d + 5)^{16}$, we replace a by v, $2b$ by w, $-3c$ by x, $2d$ by y, and 5 by z, and then apply the multinomial theorem to $(v + w + x + y + z)^{16}$. So the coefficient of $v^2w^3x^2y^5z^4$ is $\binom{16}{2,3,2,5,4} = 302{,}702{,}400$. Hence

$$\binom{16}{2,3,2,5,4}(a)^2(2b)^3(-3c)^2(2d)^5(5)^4 =$$

$$\binom{16}{2,3,2,5,4}(1)^2(2)^3(-3)^2(2)^5(5)^4(a^2b^3c^2d^5) =$$

$$435{,}891{,}456{,}000{,}000a^2b^3c^2d^5.$$

Problems

1. (a) The complete expansion of $(a + b + c + d) \cdot (e + f + g + h) \cdot (u + v + w + x + y + z)$ yields the sum of terms such as agw, cfx, and dgv. How many such terms appear in this complete expansion?

   (b) Which of the following terms do *not* appear in the complete expansion from part (a)?
   (i) afx
   (ii) bvx
   (iii) chz
   (iv) cgw
   (v) egu
   (vi) dfz

2. Determine the coefficient of $x^9 y^3$ in the expansions of:
   (a) $(x + y)^{12}$
   (b) $(x + 2y)^{12}$
   (c) $(2x - 3y)^{12}$

3. Complete the details in the proof of the multinomial theorem.

4. Determine the coefficient of:
   (a) xyz^2 in $(x + y + z)^4$
   (b) xyz^2 in $(w + x + y + z)^4$
   (c) xyz^2 in $(2x - y - z)^4$
   (d) xyz^{-2} in $(x - 2y + 3z^{-1})^4$
   (e) $w^3 x^2 yz^2$ in $(2w - x + 3y - 2z)^8$

5. Find the coefficient of $w^2 x^2 y^2 z^2$ in the expansion of:
   (a) $(w + x + y + z + 1)^{10}$
   (b) $(2w - x + 3y + z - 2)^{12}$
   (c) $(v + w - 2x + y + 5z + 3)^{12}$

6. Determine the sum of all the coefficients in the expansions of:
   (a) $(x + y)^3$
   (b) $(x + y)^{10}$
   (c) $(x + y + z)^{10}$
   (d) $(w + x + y + z)^5$
   (e) $(2s - 3t + 5u + 6v - 11w + 3x + 2y)^{10}$

7. For any positive integer n determine:

(a) $\displaystyle\sum_{i=0}^{n} \frac{1}{i!(n-i)!}$

(b) $\displaystyle\sum_{i=0}^{n} \frac{(-1)^i}{i!(n-i)!}$

8. Show that for all positive integers m and n,

$$n\binom{m+n}{m} = (m+1)\binom{m+n}{m+1}.$$

9. With n a positive integer, evaluate the sum

$$\binom{n}{0} + 2\binom{n}{1} + 2^2\binom{n}{2} + \cdots + 2^k\binom{n}{k} + \cdots + 2^n\binom{n}{n}.$$

10. For x a real number and n a positive integer, show that:

(a)

$$1 = (1+x)^n - \binom{n}{1}x^1(1+x)^{n-1} +$$
$$\binom{n}{2}x^2(1+x)^{n-2} - \cdots + (-1)^n\binom{n}{n}x^n$$

(b)

$$1 = (2+x)^n - \binom{n}{1}(x+1)(2+x)^{n-1} +$$
$$\binom{n}{2}(x+1)^2(2+x)^{n-2} - \cdots + (-1)^n\binom{n}{n}(x+1)^n$$

(c)

$$2^n = (2+x)^n - \binom{n}{1}x^1(2+x)^{n-1} +$$
$$\binom{n}{2}x^2(2+x)^{n-2} - \cdots + (-1)^n\binom{n}{n}x^n$$

11. Determine x if

$$\sum_{i=0}^{50}\binom{50}{i}8^i = x^{100}.$$

12. If a_0, a_1, a_2, a_3 is a list of four real numbers, then what is

$$\sum_{i=1}^{3}(a_i - a_{i-1})?$$

13. Given a list of $n + 1$ real numbers $a_0, a_1, a_2, \ldots, a_n$, where n is a positive integer, determine

$$\sum_{i=1}^{n}(a_i - a_{i-1}).$$

14. (a) Write a computer program (or algorithm) that lists all selections of size 2 from the objects 1, 2, 3, 4, 5, 6.

(b) Repeat part (a) for selections of size 3.

5 Combinations with Repetition

When repetitions are allowed, we've seen that for n distinct objects an arrangement of size r of these objects can be obtained in n^r ways, for an integer $r \geq 0$. We now examine the comparable problem for combinations and derive a solution by using our earlier counting principles.

Example 5.1 Seven customers at a restaurant each order one of the following food items: a cheeseburger, a hot dog, a taco, or a fish sandwich. How many different purchases are possible (from the restaurant's standpoint)?

Solution Let c, h, t, and f represent cheeseburger, hot dog, taco, and fish sandwich, respectively. We're concerned with how many of each item are purchased, not with the order in which they're purchased, so this problem is one of selections, or combinations, with repetition.

The following table lists some possible purchases in the left column and another way to represent each of those purchases in the right column.

1.	c, c, h, h, t, t, f	1. x x | x x | x x | x
2.	c, c, c, c, h, t, f	2. x x x x | x | x | x
3.	c, c, c, c, c, c, f	3. x x x x x x | | | x
4.	h, t, t, f, f, f, f	4. | x | x x | x x x x
5.	t, t, t, t, t, f, f	5. | | x x x x x | x x
6.	t, t, t, t, t, t, t	6. | | x x x x x x x |
7.	f, f, f, f, f, f, f	7. | | | x x x x x x x

For a purchase in the right column of the table, each x to the left of the first bar (|) represents a c, each x between the first and second bars represents an h, each x between the second and third bars stands for a t, and each x to the right of the third bar stands for an f. The third purchase, for example, has three consecutive bars because no one bought a hot dog or taco; the bar at the start of the fourth purchase indicates that there were no cheeseburgers in that purchase.

Hence we've established a correspondence between two collections of objects, where we already know how to count the number in one of those collections. For the representations in the right column of the table, we're counting all arrangements of 10 symbols consisting of seven x's and three |'s, so by our correspondence the number of different purchases for the right column is

$$\frac{10!}{7!\,3!} = \binom{10}{7}.$$

Note that the seven x's (one for each customer) correspond to the size of the selection and that the three bars are needed to separate the $3 + 1 = 4$ possible food items that can be chosen.

When we select, *with repetition*, r of n distinct objects, then (as in the table above) we're counting all arrangements of r x's and $n - 1$ |'s and that number is

$$\frac{(n+r-1)!}{r!(n-1)!} = \binom{n+r-1}{r}.$$

Consequently, the number of combinations of n objects taken r at a time, *with* repetition, is $C(n + r - 1, r)$.

Note that in Example 5.1, $n = 4$, $r = 7$, so it's possible for r to exceed n when repetitions are allowed.

Example 5.2   A donut shop offers 20 kinds of donuts. Assuming that there are at least a dozen of each kind in the shop, a customer can select a dozen donuts in $C(20 + 12 - 1, 12) = C(31, 12) = 141,120,525$ ways. (Here $n = 20$, $r = 12$.)

Example 5.3   President Banks has four vice presidents: (1) Adam, (2) Ben, (3) Cindy, and (4) Dawn. He wants to distribute among them $1000 in bonus checks, where each check will be written for a multiple of $100.

(a) If one or more of the vice presidents can get nothing, then President Banks is making a selection of size 10 (one for each unit of $100) from a collection of size 4 (four vice presidents), with repetition. There are $C(4 + 10 - 1, 10) = C(13, 10) = 286$ ways to do so.

(b) To avoid a mutiny, each vice president must receive at least $100. With this restriction, President Banks now must make a selection of size 6 (the remaining six units of $100) from the same collection of size 4, so the choices now number $C(4 + 6 - 1, 6) = C(9, 6) = 84$. The selection 2, 3, 3, 4, 4, 4, for example, is interpreted as follows: Adam gets nothing extra because the selection has no 1's. The one 2 in the selection indicates that Ben gets an additional $100. Cindy receives an additional $200 ($100 for each of the two 3's in the selection). Owing to the three 4's, Dawn's bonus check will total $100 + 3($100) = $400.

(c) If each vice president must get at least $100 and Dawn must get at least $500, then the number of ways to distribute the bonus checks is

$$C(3+2-1,2)+C(3+1-1,1)+C(3+0-1,0)=10=C(4+2-1,2).$$

| Dawn gets exactly $500 | Dawn gets exactly $600 | Dawn gets exactly $700 | Using the method of part (b) |

Having solved examples by using combinations with repetition, let's look at two examples that involve other counting principles as well.

Example 5.4   In how many ways can we distribute seven bananas and six oranges among four children so that each child receives at least one banana?

Solution   After giving one banana to each child, we want to count the number of ways that we can distribute the remaining three bananas among the four children. The following table shows four of these possible distributions.

1) 1, 2, 3	1) $b\,\|\,b\,\|\,b\,\|$
2) 1, 3, 3	2) $b\,\|\,\|\,b\,b\,\|$
3) 3, 4, 4	3) $\|\,\|\,b\,\|\,b\,b$
4) 4, 4, 4	4) $\|\,\|\,\|\,b\,b\,b$

For example, the second distribution in the left side of the table—namely, 1, 3, 3—indicates that we've given the first child (designated by 1) one additional banana and the third child (designated by 3) two additional bananas. The corresponding arrangement in the right column of the table represents this distribution in terms of three b's and three bars ($\|$'s). These six symbols—three of one type (the b's) and three of a second type (the bars)—can be arranged in $6!/(3!3!) = C(6, 3) = C(4 + 3 - 1, 3) = 20$ ways. (Here $n = 4$, $r = 3$.) Consequently, we have 20 ways to distribute the three additional bananas among these four children.

The following table shows the comparable situation for distributing the six oranges.

1) 1, 2, 2, 3, 3, 4	1) $o\,\|\,o\,o\,\|\,o\,o\,\|\,o$
2) 1, 2, 2, 4, 4, 4	2) $o\,\|\,o\,o\,\|\,\|\,o\,o\,o$
3) 2, 2, 2, 3, 3, 3	3) $\|\,o\,o\,o\,\|\,o\,o\,o\,\|$
4) 4, 4, 4, 4, 4, 4	4) $\|\,\|\,\|\,o\,o\,o\,o\,o\,o$

In this case we're arranging nine symbols—six of one type (the o's) and three of a second type (the bars). So we can distribute the six oranges among the four children in $9!/(6!3!) = C(9, 6) = C(4 + 6 - 1, 6) = 84$ ways. (Here $n = 4, r = 6$.)

So by the product rule there are $20 \times 84 = 1680$ ways to distribute the fruit under the stated conditions.

Example 5.5 A message made up of 12 different symbols is to be transmitted. In addition to the 12 symbols, the transmitter will also send a total of 45 spaces (blanks) between the symbols, with at least three spaces between each pair of consecutive symbols. In how many ways can such a message be sent?

Solution There are 12! ways to arrange the 12 different symbols, and for each of these arrangements there are 11 positions between the 12 symbols. Because there must be at least three spaces between successive symbols, we use up 33 of the 45 spaces and must now position the remaining 12 spaces. This is now a selection, with repetition, of size 12 (the spaces) from a collection of size 11 (the positions), and this can be done in $C(11 + 12 - 1, 12) = 646,646$ ways.

So by the product rule there are $12! \binom{22}{12} \approx 3.097 \times 10^{14}$ ways to send such messages with the required spacing.

Example 5.6 This example appears to have more to do with number theory than with combinations or arrangements, but the solution is equivalent to counting combinations with repetitions.

Determine all integer solutions of the equation

$$x_1 + x_2 + x_3 + x_4 = 7, \quad \text{where } x_i \geq 0 \text{ for all } 1 \leq i \leq 4.$$

Solution One solution of the equation is $x_1 = 3, x_2 = 3, x_3 = 0, x_4 = 1$. (This solution differs from $x_1 = 1, x_2 = 0, x_3 = 3, x_4 = 3$, even though the same four integers are used.) Let's reframe this problem as something more familiar: the solution $x_1 = 3, x_2 = 3, x_3 = 0, x_4 = 1$ means that we're distributing seven gumdrops (identical objects) among four children (distinct containers), and in this case we're giving three gumdrops to each of the first two children, nothing to the third child, and the last gumdrop to the fourth child. Continuing with this interpretation, we see that each nonnegative integer solution of the equation corresponds to a selection, with repetition, of size 7 (the *identical* gumdrops) from a collection of size 4 (the *distinct* children), so there are $C(4 + 7 - 1, 7) = 120$ solutions.

At this point we can see that the following values are equivalent:

(a) The number of integer solutions of the equation

$$x_1 + x_2 + \cdots + x_n = r, \qquad x_i \geq 0, \qquad 1 \leq i \leq n.$$

(b) The number of selections, with repetition, of size r from a collection of size n.

(c) The number of ways that r identical objects can be distributed among n distinct containers.

Note that, in terms of distributions, part (c) is valid only when the r objects being distributed are *identical* and the n containers are distinct. If both the r objects and the n containers are distinct, then by the product rule we can select any of the n containers for each one of the objects and get n^r distributions.

Example 5.7   In how many ways can 10 (identical) black marbles be distributed among six distinct containers?

Solution   Solving this problem is equivalent to finding the number of nonnegative integer solutions of the equation $x_1 + x_2 + \cdots + x_6 = 10$, which is the number of selections of size 10, with repetition, from a collection of size 6. Hence the answer is $C(6 + 10 - 1, 10) = 3003$.

Example 5.8   From the preceding example we know that there are 3003 nonnegative integer solutions of the equation $x_1 + x_2 + \cdots + x_6 = 10$. How many such solutions are there to the inequality $x_1 + x_2 + \cdots + x_6 < 10$?

Solution   One approach is to determine the number of such solutions to $x_1 + x_2 + \cdots + x_6 = k$, where k is an integer and $0 \leq k \leq 9$. This approach is feasible in this example, but is impractical if 10 is replaced by, say,

100 or another large number. It's possible to establish a combinatorial identity that can be used to solve the problem by using this approach, but that area of combinatorics is beyond the scope of this book. For now we'll transform the problem by noting the correspondence between the nonnegative integer solutions of

$$x_1 + x_2 + \cdots + x_6 < 10 \qquad (1)$$

and the integer solutions of

$$x_1 + x_2 + \cdots + x_6 + x_7 = 10, \quad x_i \geq 0, \quad 1 \leq i \leq 6, \quad x_7 > 0. \qquad (2)$$

The number of solutions of equation (2) is the same as the number of nonnegative integer solutions of $y_1 + y_2 + \cdots + y_6 + y_7 = 9$, where $y_i = x_i$, for $1 \leq i \leq 6$, and $y_7 = x_7 - 1$. This number is $C(7 + 9 - 1, 9) = 5005$.

Example 5.9 This example involves binomial and multinomial expansions (Chapter 4).

In the binomial expansion for $(x + y)^n$, each term is of the form $\binom{n}{k}$ $x^k y^{n-k}$, so the total number of terms in the expansion is the number of nonnegative integer solutions of $n_1 + n_2 = n$ (n_1 is the exponent for x and n_2 is the exponent for y). This number is $C(2 + n - 1, n) = n + 1$.

This argument might strike you as a bit roundabout, and you might prefer instead to accept the result on the basis of your experience in expanding $(x + y)^n$ for various small values of n. But experience isn't always adequate to find a general principle, and here it would prove worthless if we wanted to know how many terms are in the expansion of $(w + x + y + z)^{10}$.

Each distinct term here is of the form

$$\binom{10}{n_1, n_2, n_3, n_4} w^{n_1} x^{n_2} y^{n_3} z^{n_4},$$

where $n_i \geq 0$, for $1 \leq i \leq 4$, and $n_1 + n_2 + n_3 + n_4 = 10$. This last equation can be solved in $C(4 + 10 - 1, 10) = 286$ ways, so there are 286 terms in the expansion of $(w + x + y + z)^{10}$.

Example 5.10 This example involves the binomial expansion and part (a) of Corollary 4.1 in Chapter 4.

(a) Let's determine all the different ways in which we can write the number 4 as a sum of positive integers, where the order of the summands is relevant. These representations, called the **compositions** of 4, are:

(1) 4
(2) 3 + 1
(3) 1 + 3
(4) 2 + 2
(5) 2 + 1 + 1
(6) 1 + 2 + 1
(7) 1 + 1 + 2
(8) 1 + 1 + 1 + 1

The number 4 has eight compositions in total, including the sum consisting of only one summand—namely, 4. (If we don't care about the order of the summands, then representations (2) and (3) are the same, as are representations (5), (6), and (7). In this case there are five **partitions** of the number 4—namely, 4; 3 + 1; 2 + 2; 2 + 1 + 1; and 1 + 1 + 1 + 1.)

(b) Now suppose that we want to *count* the compositions of the number 7. Here we *don't* want to list all the possibilities—which include 7; 6 + 1; 1 + 6; 5 + 2; 1 + 2 + 4; 2 + 4 + 1; and 3 + 1 + 2 + 1. To count all these compositions, let's consider the number of possible summands.

(i) For one summand there's only one composition—namely, 7.

(ii) For two (positive) summands, we count the number of *positive* integer solutions for

$$w_1 + w_2 = 7, \quad \text{where } w_1, w_2 > 0,$$

which equals the number of *nonnegative* integer solutions for

$$x_1 + x_2 = 5, \quad \text{where } x_1, x_2 \geq 0.$$

The number of such solutions is

$$\binom{2+5-1}{5} = \binom{6}{5}.$$

(iii) For three (positive) summands, we count the number of *positive* integer solutions for

$$y_1 + y_2 + y_3 = 7,$$

which equals the number of *nonnegative* integer solutions for

$$z_1 + z_2 + z_3 = 4.$$

The number of such solutions is

$$\binom{3+4-1}{4} = \binom{6}{4}.$$

The following list summarizes cases (i), (ii), (iii), and the other four cases, where n = the number of summands in a composition of 7 and m = the number of compositions of 7 with n summands. (Note for case (i) that $1 = \binom{6}{6}$.)

(i) $n = 1$, $m = C(6, 6)$
(ii) $n = 2$, $m = C(6, 5)$
(iii) $n = 3$, $m = C(6, 4)$
(iv) $n = 4$, $m = C(6, 3)$
(v) $n = 5$, $m = C(6, 2)$
(vi) $n = 6$, $m = C(6, 1)$
(vii) $n = 7$, $m = C(6, 0)$

Consequently, the m values listed above tell us that the (total) number of compositions of 7 is

$$\binom{6}{6} + \binom{6}{5} + \binom{6}{4} + \binom{6}{3} + \binom{6}{2} + \binom{6}{1} + \binom{6}{0} = \sum_{k=0}^{6} \binom{6}{k}.$$

From part (a) of Corollary 4.1 this reduces to 2^6.

In general, for each positive integer m, there are

$$\sum_{k=0}^{m-1}\binom{m-1}{k} = 2^{m-1}$$

compositions.

Example 5.11 From the preceding example we know that there are $2^{12-1} = 2^{11} = 2048$ compositions of 12. To count those compositions where each summand is even, we consider compositions such as

$2 + 4 + 6 = 2(1 + 2 + 3)$,
$8 + 2 + 2 = 2(4 + 1 + 1)$,
$2 + 8 + 2 = 2(1 + 4 + 1)$,
$6 + 6 = 2(3 + 3)$.

In each of these four examples, the parenthesized expression is a composition of 6. This observation tells us that the number of compositions of 12 where each summand is even equals the number of (all) compositions of 6, which is $2^{6-1} = 2^5 = 32$.

Example 5.12 This example illustrates a programming application.

Consider the following program snippet, where i, j, and k are integer variables.

```
for i := 1 to 20 do
  for j := 1 to i do
    for k := 1 to j do
      print (i * j + k)
```

How many times is the print statement executed in this program snippet?

Solution Among the possible choices for i, j, and k (in the order i-first, j-second, k-third) that will lead to execution of the print statement are (1) 1, 1, 1; (2) 2, 1, 1; (3) 15, 10, 1; and (4) 15, 10, 7. Note that $i = 10, j = 12, k = 5$ isn't a valid selection because $j = 12 > 10 = i$ violates the condition set forth in the second for loop. Each of the above four selections where the print statement is executed satisfies the condition $1 \le k \le j \le i \le 20$. In fact, any selection a, b, c ($a \le b \le c$) of size 3, with repetitions allowed, from the list 1, 2, 3, …, 20 results in one of the correct selections: here, $k = a, j = b, i = c$. Consequently the print statement is executed

$$\binom{20+3-1}{3} = \binom{22}{3} = 1540 \text{ times.}$$

If there had been $r\ (\geq 1)$ for loops instead of three, then the print statement would have been executed $\binom{20+r-1}{r}$ times.

Example 5.13 In this example we use a program to derive an important summation formula. In the following program snippet, the variables i, j, n, and *counter* are integer variables. We assume that the value of n has been set prior to executing the snippet.

```
counter := 0
for i := 1 to n do
  for j := 1 to i do
    counter := counter + 1
```

From the results in Example 5.12, after this snippet is executed the value of the variable *counter* will be $\binom{n+2-1}{2} = \binom{n+1}{2}$. This value is also the number of times that the statement

(*)   counter := counter + 1

is executed.

This result can also be reasoned as follows: when i is assigned a value of 1, then j varies from 1 to 1 and (*) is executed once; when $i := 2$, then j varies from 1 to 2 and (*) is executed twice; when $i := 3$, then j varies from 1 to 3 and (*) is executed three times. In general, for $1 \leq k \leq n$, when $i := k$, then j varies from 1 to k and (*) is executed k times. In total, the variable *counter* is incremented (and the statement (*) is executed) $1 + 2 + 3 + \cdots + n$ times.

Consequently,

$$\sum_{i=1}^{n} i = 1 + 2 + 3 + \cdots + n = \binom{n+1}{2} = \frac{n(n+1)}{2}.$$

The derivation of this summation formula, obtained by counting the same result in two different ways, constitutes a combinatorial proof.

Example 5.14 This example introduces the idea of a run, a concept that arises in statistics—in particular, in the detecting of trends in a statistical process.

The counter at a saloon has 15 bar stools that are occupied as follows:

$$O\ O\ E\ O\ O\ O\ O\ E\ E\ E\ O\ O\ O\ E\ O,$$

where O indicates an occupied stool and E an empty one. (Here we're not concerned with the distinct occupants of the stools, only whether a stool is occupied.) The current occupancy of the 15 stools determines seven runs, as shown:

$$\underbrace{OO}_{\text{Run}}\ \underbrace{E}_{\text{Run}}\ \underbrace{OOOO}_{\text{Run}}\ \underbrace{EEE}_{\text{Run}}\ \underbrace{OOO}_{\text{Run}}\ \underbrace{E}_{\text{Run}}\ \underbrace{O}_{\text{Run}}.$$

In general, a **run** is a consecutive list of identical entities that are preceded and followed by different entities or no entities at all.

A second way in which five E's and 10 O's can be arranged to provide seven runs is

$$E\ O\ O\ O\ E\ E\ O\ O\ E\ O\ O\ O\ O\ O\ E.$$

We want to count the total number of ways that five E's and 10 O's can determine seven runs. If the first run starts with an E, then there must be four runs of E's and three runs of O's. Consequently, the last run must end with an E.

Let x_1 count the number of E's in the first run, x_2 the number of O's in the second run, x_3 the number of E's in the third run,..., and x_7 the number of E's in the seventh run. We want to find the number of integer solutions for

$$x_1 + x_3 + x_5 + x_7 = 5, \quad x_1, x_3, x_5, x_7 > 0 \qquad (3)$$

and

$$x_2 + x_4 + x_6 = 10, \quad x_2, x_4, x_6 > 0. \qquad (4)$$

The number of integer solutions for equation (3) equals the number of integer solutions for

$$y_1 + y_3 + y_5 + y_7 = 1, \quad y_1, y_3, y_5, y_7 \geq 0.$$

This number is $\binom{4+1-1}{1} = \binom{4}{1} = 4$. Similarly, for equation (4) the number

of solutions is $\binom{3+7-1}{7} = \binom{9}{7} = 36$. Consequently, by the product rule

there are $4 \cdot 36 = 144$ arrangements of five E's and 10 O's that determine seven runs, the first run starting with E.

The seven runs can also have the first run starting with an O and the last run ending with an O. So now let w_1 count the number of O's in the first run, w_2 the number of E's in the second run, w_3 the number of O's in the third run,..., and w_7 the number of O's in the seventh run. Here we want the number of integer solutions for

$$w_1 + w_3 + w_5 + w_7 = 10, \quad w_1, w_3, w_5, w_7 > 0.$$

and

$$w_2 + w_4 + w_6 = 5, \quad w_2, w_4, w_6 > 0.$$

Reasoning as above, the number of ways to arrange five E's and 10 O's, resulting in seven runs where the first run starts with an O, is

$$\binom{4+6-1}{6}\binom{3+2-1}{2} = \binom{9}{6}\binom{4}{2} = 504.$$

So by the sum rule the five E's and 10 O's can be arranged in $144 + 504 = 648$ ways to produce seven runs.

Problems

1. In how many ways can 10 (identical) oranges be distributed among five children if:
   (a) there are no restrictions?
   (b) each child gets at least one orange?
   (c) the oldest child gets at least two oranges?

2. In how many ways can 15 (identical) gumdrops be distributed among five children so that the youngest gets only one or two of them?

3. Determine how many ways that 20 coins can be selected from four large containers filled with pennies, nickels, dimes, and quarters. (Each container is filled with only one type of coin.)

4. An ice cream store has 31 flavors available. In how many ways can we order a dozen ice cream cones if:
   (a) we don't want the same flavor more than once?
   (b) a flavor can be ordered as many as 12 times?
   (c) a flavor can be ordered no more than 11 times?

5. (a) In how many ways can we select five coins from a collection of 10 consisting of one penny, one nickel, one dime, one quarter, one half-dollar, and five (identical) silver dollars?

   (b) In how many ways can we select n objects from a collection of size $2n$ that consists of n distinct and n identical objects?

6. Answer Example 5.5 where the 12 symbols being transmitted are four A's, four B's, and four C's.

7. Determine the number of integer solutions of
$$x_1 + x_2 + x_3 + x_4 = 32,$$
   where
   (a) $x_i \geq 0, \quad 1 \leq i \leq 4$
   (b) $x_i > 0, \quad 1 \leq i \leq 4$
   (c) $x_1, x_2 \geq 5, \quad x_3, x_4 \geq 7$
   (d) $x_i \geq 8, \quad 1 \leq i \leq 4$
   (e) $x_i \geq -2, \quad 1 \leq i \leq 4$
   (f) $x_1, x_2, x_3 > 0, \quad 0 < x_4 \leq 25$

8. In how many ways can a parent distribute eight chocolate donuts and seven jelly donuts among three children if each child wants at least one donut of each kind?

9. Sally has two dozen each of n different colored beads. If she can select 20 beads (with repetitions of colors allowed) in 230,230 ways, then what is the value of n?

10. In how many ways can Jane toss 100 (identical) dice so that at least three of each type of face will be showing?

11. Two n-digit integers (leading zeros allowed) are considered to be equivalent if one is a rearrangement of the other. For example, 12033, 20331, and 01332 are considered to be equivalent five-digit integers.

    (a) How many five-digit integers are not equivalent?

    (b) If the digits 1, 3, and 7 can appear at most once, then how many nonequivalent five-digit integers are there?

12. Determine the number of integer solutions of
$$x_1 + x_2 + x_3 + x_4 + x_5 < 40,$$
where
    (a) $x_i \geq 0, \quad 1 \leq i \leq 5$
    (b) $x_i \geq -3, \quad 1 \leq i \leq 5$

13. In how many ways can we distribute eight identical black balls into four distinct containers so that:
    (a) no container is left empty?
    (b) the fourth container has an odd number of balls in it?

14. (a) Find the coefficient of $v^2 w^4 xz$ in the expansion of $(3v + 2w + x + y + z)^8$.

    (b) How many distinct terms arise in the expansion in part (a)?

15. In how many ways can we place 24 different books on four shelves so that there is at least one book on each shelf? (For any of these arrangements consider the books on each shelf to be placed one next to the other, with the first book at the left of the shelf.)

16. How many ways are there to place 12 marbles of the same size in five distinct jars if:
    (a) the marbles are all black?
    (b) each marble is a different color?

17. In the following program snippet, i, j, k, and m are integer variables. How many times does the print statement execute?

```
for i := 1 to 20 do
  for j := 1 to i do
    for k := 1 to j do
      for m := 1 to k do
        print ((i * j) + (k * m))
```

18. In the following program snippet, i, j, k, and *counter* are integer variables. What is the final value of *counter* after the snippet executes?

```
counter := 10
for i := 1 to 15 do
  for j := i to 15 do
    for k := j to 15 do
      counter := counter + 1
```

19. In the following program snippet, i, j, k, *increment*, and *sum* are integer variables. What is the final value of *sum* after the snippet executes?

```
increment := 0
sum := 0
for i := 1 to 10 do
  for j := 1 to i do
    for k := 1 to j do
      begin
        increment := increment + 1
        sum := sum + increment
      end
```

20. Given positive integers m, n with $m \geq n$, show that the number of ways to distribute m identical objects into n distinct containers with no container left empty is $C(m - 1, m - n) = C(m - 1, n - 1)$.

21. Write a computer program (or algorithm) to list the integer solutions for
    (a) $x_1 + x_2 + x_3 = 10$, $\quad x_i \geq 0$, $\quad 1 \leq i \leq 3$
    (b) $x_1 + x_2 + x_3 + x_4 = 4$, $\quad x_i \geq -2$, $\quad 1 \leq i \leq 4$

22. Of the 2^{19} compositions of 20, how many have every summand even?

23. Let n, m, k be positive integers where $n = mk$. How many compositions of n have every summand a multiple of k?

6

Summary and Solutions

This book introduced the fundamentals for counting combinations, permutations, and arrangements in a large variety of problems. Solving nontrivial counting problems requires breaking them into parts, solving each subproblem (possibly by using a variety of different formulas), and then assembling the solved components to arrive at the final answer. This stepwise process is akin to *top-down programming*, where a programmer develops an algorithm for solving a hard problem first by considering major subproblems that must be solved. These subproblems are then further refined—subdivided into more easily workable programming tasks. Each level of refinement improves on the clarity, precision, and completeness of the algorithm until it can be readily translated into code.

Given a collection of n distinct objects, the following formulas count the number of ways to select, or order, with or without repetitions, r of these n objects.

Order Relevant	Repetitions Allowed	Type of Result	Formula
Yes	No	Permutation	$P(n,r) = n!/(n-r)!$, $0 \leq r \leq n$
Yes	Yes	Arrangement	n^r, $n,r \geq 0$
No	No	Combination	$C(n,r) = n!/[r!(n-r)!] = \binom{n}{r}$, $0 \leq r \leq n$
No	Yes	Combination with repetition	$\binom{n+r-1}{r}$, $n,r \geq 0$

Chapter 2

1. (a) By the sum rule there are $8 + 5 = 13$ possibilities for the eventual winner.

   (b) Because there are eight party A candidates and five party B candidates, by the product rule we have $8 \times 5 = 40$ possible pairs of opposing candidates.

   (c) The sum rule in part (a); the product rule in part (b).

2. By the product rule there are $5 \times 5 \times 5 \times 5 \times 5 \times 5 = 5^6$ license plates where the first two symbols are vowels and the last four are even digits.

3. (a) By the product rule there are $4 \times 12 \times 3 \times 2 = 288$ distinct automobiles that can be manufactured.

   (b) Of these, $4 \times 1 \times 3 \times 2 = 24$ are blue.

4. (a) By the product rule there are $10 \times 9 \times 8 \times 7 = P(10, 4) = 5040$ possible slates.

   (b) (i) There are $3 \times 9 \times 8 \times 7 = 1512$ slates where an engineer is nominated for president. (ii) The number of slates with exactly one engineer appearing is $4 \times [3 \times 7 \times 6 \times 5] = 2520$. (iii) There are $7 \times 6 \times 5 \times 4 = 840$ slates where no engineer is nominated for any of the four offices. Consequently, $5040 - 840 = 4200$ slates include at least one engineer.

5. Based on the eyewitness testimony, by the product rule there are $2 \times 2 \times 1 \times 10 \times 10 \times 2 = 800$ different license plates to investigate.

6. (a) We have permutations of 30 objects (the runners) taken 8 (the first eight finishing positions) at a time. So the trophies can be awarded in $P(30, 8) = 30!/22!$ ways.

   (b) Chris and David can finish among the top three runners in 6 ways. For each of these 6 ways, there are $P(28, 6)$ ways for the other 6 finishers (in the top 8) to finish the race. By the product rule there are $6 \cdot P(28, 6)$ ways to award the trophies with these two runners among the top three.

7. By the product rule there are 2^9 possibilities.

8. By the product rule there are (a) 12! ways to process the programs if there are no restrictions; (b) (4!)(8!) ways so that the four higher-priority programs are processed first; and (c) (4!)(5!)(3!) ways where the four top-priority programs are processed first and the three least-priority programs are processed last.

9. (a) (14)(12) = 168.
   (b) (14)(12)(6)(18) = 18,144.
   (c) (8)(18)(6)(3)(14)(12)(14)(12) = 73,156,608.

10. Consider one such arrangement: say we have three books on one shelf and 12 on the other. This can be done in 15! ways. In fact, for any subdivision (resulting in two nonempty shelves) of the 15 books we get 15! ways to arrange the books on the two shelves. Because there are 14 ways to subdivide the books so that each shelf has at least one book, the total number of ways in which Alice can arrange her books in this manner is (14)(15!).

11. (a) There are four roads from town A to town B and three roads from town B to town C, so by the product rule there are $4 \times 3 = 12$ roads from A to C that pass through B. Because there are two roads from A to C directly, there are $12 + 2 = 14$ ways in which you can travel from A to C.

    (b) Using the result from part (a), together with the product rule, there are $14 \times 14 = 196$ different round trips (from A to C and then back to A).

    (c) Here there are $14 \times 13 = 182$ round trips.

12. act, atc, cat, cta, tac, tca.

13. (a) $8! = P(8, 8)$.
    (b) 7!, 6!.

14. (a) $P(7, 2) = 7!/(7 - 2)! = 7!/5! = (7)(6) = 42.$
    (b) $P(8, 4) = 8!/(8 - 4)! = 8!/4! = (8)(7)(6)(5) = 1680.$
    (c) $P(10, 7) = 10!/(10 - 7)! = 10!/3! = (10)(9)(8)(7)(6)(5)(4) = 604,800.$
    (d) $P(12, 3) = 12!/(12 - 3)! = 12!/9! = (12)(11)(10) = 1320.$

15. Here we must place a, b, c, d in the positions denoted by X: e X e X e X e X e. By the product rule there are 4! ways to do so.

16. (a) With repetitions allowed there are 40^{25} distinct messages.

    (b) By the product rule there are $40 \times 30 \times 30 \times \cdots \times 30 \times 30 \times 40 = (40^2)(30^{23})$ messages.

17. Class A: $(2^7 - 2)(2^{24} - 2) = 2{,}113{,}928{,}964$.
    Class B: $2^{14}(2^{16} - 2) = 1{,}073{,}709{,}056$.
    Class C: $2^{21}(2^8 - 2) = 532{,}676{,}608$.

18. By the product rule there are $(7)(4)(3)(6) = 504$ ways for Daisy to configure her computer system.

19. (a) $7! = 5040$.
    (b) $4 \times 3 \times 3 \times 2 \times 2 \times 1 \times 1 = (4!)(3!) = 144$.
    (c) $(3!)(5)(4!) = 720$.
    (d) $(3!)(4!)(2) = 288$.

20. (a) Because there are three A's, there are $8!/3! = 6720$ arrangements.

    (b) The six symbols D, T, B, S, E, AAA can be arranged in $6! = 720$ ways.

21. (a) $12!/(3!2!2!2!)$.

    (b) $[11!/(3!2!2!2!)]$ (for AG) + $[11!/(3!2!2!2!)]$ (for GA).

    (c) Consider one case where all the vowels are adjacent: S, C, L, G, C, L, OIOOIA. These seven symbols can be arranged in $(7!)/(2!2!)$ ways. Because O, O, O, I, I, A can be arranged in $(6!)/(3!2!)$ ways, the number of arrangements with all the vowels adjacent is $[7!/(2!2!)][6!/(3!2!)]$.

22. Case 1, leading digit is 5: $(6!)/(2!)$.
    Case 2, leading digit is 6: $(6!)/(2!)^2$.
    Case 3, leading digit is 7: $(6!)/(2!)^2$.

    In total there are $[(6!)/(2!)][1 + (1/2) + (1/2)] = 6! = 720$ such positive integers n.

23. Here the solution is the number of ways that we can arrange 12 objects: 4 the first type, 3 of the second type, 2 of the third type, and 3 of the fourth type. There are $12!/(4!3!2!3!) = 277,200$ ways.

24. $P(n + 1, r) = (n + 1)!/(n + 1 - r)! = [(n + 1)/(n + 1 - r)] \cdot [n!/(n - r)!] = [(n + 1)/(n + 1 - r)]P(n, r)$.

25. (a) $n = 10$.

(b) $n = 5$.

(c) $2n!/(n - 2)! + 50 = (2n)!/(2n - 2)! \Rightarrow 2n(n - 1) + 50 = (2n)(2n - 1) \Rightarrow n^2 = 25 \Rightarrow n = 5$.

26. Any such path from $(0, 0)$ to $(7, 7)$ or from $(2, 7)$ to $(9, 14)$ is an arrangement of 7 R's and 7 U's; there are $(14!)/(7!7!)$ such arrangements. In general, for nonnegative integers m, n, and any real numbers a, b, the number of such paths from (a, b) to $(a + m, b + n)$ is $(m + n)!/(m!n!)$.

27. (a) Each path consists of 2 H's, 1 V, and 7 A's. There are $10!/(2!1!7!)$ ways to arrange these 10 letters and this is the number of paths.

(b) $10!/(7!1!2!)$.

(c) If a, b, c are any real numbers and m, n, p are nonnegative integers, then the number of paths from (a, b, c) to $(a + m, b + n, c + p)$ is $(m + n + p)!/(m!n!p!)$.

28. (a) The for loop for i is executed 12 times, while those for j and k are executed $10 - 5 + 1 = 6$ and $15 - 8 + 1 = 8$ times, respectively. Consequently, following the execution of the given program snippet, the value of *counter* is $0 + 12(1) + 6(2) + 8(3) = 48$.

(b) Here we have three tasks: T_1, T_2, and T_3. Task T_1 takes place each time we traverse the instructions in the i loop. Similarly, tasks T_2 and T_3 take place during each iteration of the j and k loops, respectively. The final value for the integer variable *counter* follows by the sum rule.

29. (a) and (b) By the product rule the `print` statement is executed $12 \times 6 \times 8 = 576$ times.

30. (a) For five letters there are $26 \times 26 \times 26 \times 1 \times 1 = 26^3$ palindromes for six letters.

(b) When letters can't appear more than two times, there are $26 \times 25 \times 24 = 15{,}600$ palindromes for either five or six letters.

31. (a) By the product rule there are $9 \times 9 \times 8 \times 7 \times 6 \times 5 = 136{,}080$ six-digit integers with no leading zeros and no repeated digit.

(b) When digits can he repeated there are 9×10^5 such six-digit integers.

(c) (i) (a) $(9 \times 8 \times 7 \times 6 \times 5 \times 1)$ (for the integers ending in 0) + $(8 \times 8 \times 7 \times 6 \times 5 \times 4)$ (for the integers ending in 2, 4, 6, or 8) = 68,800. (b) When the digits can be repeated there are $9 \times 10 \times 10 \times 10 \times 10 \times 5 = 450{,}000$ six-digit even integers.

(c) (ii) (a) $(9 \times 8 \times 7 \times 6 \times 5 \times 1)$ (for the integers ending in 0) + $(8 \times 8 \times 7 \times 6 \times 5 \times 1)$ (for the integers ending in 5) = 28,560. (b) $9 \times 10 \times 10 \times 10 \times 10 \times 2 = 180{,}000$.

(c) (iii) We use the fact that an integer is divisible by 4 if and only if the integer formed by the last two digits is divisible by 4. (a) $(8 \times 7 \times 6 \times 5 \times 6)$ (last two digits are 04, 08, 20, 40, 60, or 80) + $(7 \times 7 \times 6 \times 5 \times 16)$ (last two digits are 12, 16, 24, 28, 32, 36, 48, 52, 56, 64, 68, 72, 76, 84, 92, or 96) = 33,600. (b) $9 \times 10 \times 10 \times 10 \times 25 = 225{,}000$.

32. (a) For positive integers n, k, where $n = 3k$, $n!/(3!)^k$ is the number of ways to arrange the n objects $x_1, x_1, x_1, x_2, x_2, x_2, \ldots, x_k, x_k, x_k$. This number must be an integer.

(b) If n, k are positive integers where $n = mk$, then $n!/(m!)^k$ is an integer.

33. (a) With 2 choices per question there are $2^{10} = 1024$ ways to answer the test.

(b) Now there are 3 choices per question and 3^{10} ways.

34. (4!/2!) (no 7's) + (4!) (one 7 and one 3) + (2)(4!/2!) (one 7 and two 3's) + (4!/2!) (two 7's and no 3's) + (2)(4!/2!) (two 7's and one 3) + (4!/(2!2!)) (two 7's and two 3's). The total gives us 102 such four-digit integers.

35. (a) 6!.

(b) Let A, B denote the two people who insist on sitting next to each other. Then there are 5! (A to the right of B) + 5! (B to the right of A) = 2(5!) seating arrangements.

36. (a) Locate A. There are two cases to consider: (1) There is a person to the left of A on the same side of the table. There are 7! such seating arrangements. (2) There is a person to the right of A on the same side of the table. This gives 7! more arrangements. So there are 2(7!) possibilities.

(b) 7200.

Chapter 3

1. $\binom{6}{2} = 6!/[2!(6-2)!] = 6!/(2!4!) = (6)(5)/2 = 15.$

ab	bc	ce
ac	bd	cf
ad	be	de
ae	bf	df
af	cd	ef

2. Order is unimportant here, so Jane can make her selection in $\binom{12}{5} =$ 792 ways.

3. (a) $C(10, 4) = 10!/(4!6!) = (10)(9)(8)(7)/(4)(3)(2)(1) = 210.$

(b) $\binom{12}{7} = 12!/(7!5!) = (12)(11)(10)(9)(8)/(5)(4)(3)(2)(1) = 792.$

(c) $C(14, 12) = 14!/(12!2!) = (14)(13)/(2)(1) = 91.$

(d) $\binom{15}{10} = 15!/(10!5!) = (15)(14)(13)(12)(11)/(5)(4)(3)(2)(1) = 3003.$

4. (a) $2^6 - 1 = 63$.

   (b) $\dbinom{6}{3} = 20$.

   (c) $\dbinom{6}{2} + \dbinom{6}{4} + \dbinom{6}{6} = 31$.

5. (a) $P(5, 3) = 5!/(5 - 3)! = 5!/2! = (5)(4)(3) = 60$ permutations of size 3.

   (b) $C(5, 3) = 5!/[3!(5 - 3)!] = 5!/(3!2!) = 10$ combinations of size 3. They are:

afm	art
afr	fmr
aft	fmt
amr	frt
amt	mrt

6. $\dbinom{n}{2} + \dbinom{n-1}{2} = \frac{1}{2}(n)(n-1) + \frac{1}{2}(n-1)(n-2)$

   $= \frac{1}{2}(n-1)[n + (n-2)]$

   $= \frac{1}{2}(n-1)(2n-2)$

   $= (n-1)^2$.

7. (a) $\dbinom{20}{12}$.

   (b) $\dbinom{10}{6}\dbinom{10}{6}$.

   (c) $\dbinom{10}{2}\dbinom{10}{10}$ (2 women) $+ \dbinom{10}{4}\dbinom{10}{8}$ (4 women) $+ \cdots +$

   $\dbinom{10}{10}\dbinom{10}{2}$ (10 women) $= \sum_{i=1}^{5} \dbinom{10}{2i}\dbinom{10}{12-2i}$.

   (d) $\dbinom{10}{7}\dbinom{10}{5}$ (7 women) $+ \dbinom{10}{8}\dbinom{10}{4}$ (8 women) $+$

   $\dbinom{10}{9}\dbinom{10}{3}$ (9 women) $+ \dbinom{10}{10}\dbinom{10}{2}$ (10 women)

   $= \sum_{i=7}^{10} \dbinom{10}{i}\dbinom{10}{12-i}$.

(e) $\displaystyle\sum_{i=8}^{10}\binom{10}{i}\binom{10}{12-i}$.

8. (a) $\binom{4}{1}\binom{13}{5}$.

   (b) $\binom{4}{4}\binom{48}{1}$.

   (c) $\binom{13}{1}\binom{4}{4}\binom{48}{1}$.

   (d) $\binom{4}{3}\binom{4}{2}$.

   (e) $\binom{4}{3}\binom{12}{1}\binom{4}{2}$.

   (f) $\binom{13}{1}\binom{4}{3}\binom{12}{1}\binom{4}{2} = 3744$.

   (g) $\binom{13}{1}\binom{4}{3}\binom{48}{1}\binom{44}{1}\cdot\dfrac{1}{2}$.

   Division by 2 is needed because no distinction is made for the order in which the other two cards are drawn. The result equals

   $$54{,}912 = \binom{13}{1}\binom{4}{3}\binom{48}{2} - 3744 = \binom{13}{1}\binom{4}{3}\binom{12}{2}\binom{4}{1}\binom{4}{1}.$$

   (h) $\binom{13}{2}\binom{4}{2}\binom{4}{2}\binom{44}{1}$.

9. (a) $\binom{8}{2}$.

   (b) $\binom{8}{4}$.

   (c) $\binom{8}{6}$.

   (d) $\binom{8}{6}+\binom{8}{7}+\binom{8}{8}$.

10. (a) $\binom{12}{5}$.

 (b) $\binom{10}{3}$.

11. (a) $\binom{10}{7} = 120$.

 (b) $\binom{8}{5} = 56$.

 (c) $\binom{6}{4}\binom{4}{3}$ (four of the first six) +

 $\binom{6}{5}\binom{4}{2}$ (five of the first six) +

 $\binom{6}{6}\binom{4}{1}$ (all of the first six)

 $= (15)(4) + (6)(6) + (1)(4) = 100$.

12. (a) The first three books can be selected in $\binom{12}{3}$ ways, the next three in $\binom{9}{3}$ ways, the third set of three in $\binom{6}{3}$ ways, and the fourth set in $\binom{3}{3}$ ways. Consequently, the books can be distributed in

 $\binom{12}{3}\binom{9}{3}\binom{6}{3}\binom{3}{3} = (12!)/[(3!)^4]$ ways.

 (b) $\binom{12}{4}\binom{8}{4}\binom{4}{2}\binom{2}{2} = (12!)/[(4!)^2(2!)^2]$.

13. The letters M, I, I, I, P, P, I can be arranged in $[7!/(4!)(2!)]$ ways. Each arrangement provides eight positions (one at the start of the arrangement, one at the end, and six between letters) for placing four nonconsecutive S's. Four of these positions can be selected in $\binom{8}{4}$ ways. So the total number of arrangements is $\binom{8}{4}[7!/(4!)(2!)]$.

14. $\binom{n}{11} = 12{,}376$ when $n = 17$.

15. (a) Two distinct points determine a line. With 15 points, no three collinear, there are $\binom{15}{2}$ possible lines.

(b) There are $\binom{25}{3}$ possible triangles or planes, and $\binom{25}{4}$ possible tetrahedra.

16. (a) $\sum\limits_{i=1}^{6}(i^2+1) = (1^2+1) + (2^2+1) + (3^2+1) + (4^2+1) + (5^2+1) +$
$(6^2+1) = 2 + 5 + 10 + 17 + 26 + 37 = 97.$

(b) $\sum\limits_{j=-2}^{2}(j^3-1) = [(-2)^3-1] + [(-1)^3-1] + (0^3-1) + (1^3-1) + (2^3-1) =$
$-9 - 2 - 1 + 0 + 7 = -5.$

(c) $\sum\limits_{i=0}^{10}[1+(-1)^i] = 2 + 0 + 2 + 0 + 2 + 0 + 2 + 0 + 2 + 0 + 2 = 12.$

(d) $\sum\limits_{k=n}^{2n}(-1)^k = [(-1)^n + (-1)^{n+1}] + [(-1)^{n+2} + (-1)^{n+3}] + \cdots +$
$[(-1)^{2n-1} + (-1)^{2n}] = 0 + 0 + \cdots + 0 = 0.$

(e) $\sum\limits_{i=1}^{6}i(-1)^i = -1 + 2 - 3 + 4 - 5 + 6 = 3.$

17. (a) $\sum\limits_{k=2}^{n}\dfrac{1}{k!}.$

(b) $\sum\limits_{i=1}^{7}i^2.$

(c) $\sum\limits_{j=1}^{7}(-1)^{j-1}j^3 = \sum\limits_{k=1}^{7}(-1)^{k+1}k^3.$

(d) $\sum\limits_{i=0}^{n}\dfrac{i+1}{n+i}.$

(e) $\sum\limits_{i=0}^{n}(-1)^i\left[\dfrac{n+i}{(2i)!}\right].$

18. (a) $10!/(4!3!3!)$.

(b) $\binom{10}{8}2^2 + \binom{10}{9}2 + \binom{10}{10}$.

(c) $\binom{10}{4}$ (four 1's, six 0's) +

$\binom{10}{2}\binom{8}{1}$ (two 1's, one 2, seven 0's) +

$\binom{10}{2}$ (two 2's, eight 0's).

19. (a) $\binom{10}{3}$ (three 1's, seven 0's) +

$\binom{10}{1}\binom{9}{1}$ (one 1, one 2, eight 0's) +

$\binom{10}{1}$ (one 3, nine 0's)

$= 220$.

(b) $\binom{10}{4} + \binom{10}{2} + \binom{10}{1}\binom{9}{2} + \binom{10}{1}\binom{9}{1} = 705$.

(c) $2^{10} \cdot \sum_{i=0}^{5}\binom{10}{2i}$.

Select an even number of positions for 0, 2. This is done in $\binom{10}{2i}$ ways for $0 \le i \le 5$. Then for the $2i$ positions selected there are two choices; for the $10 - 2i$ remaining positions there are also two choices—namely, 1, 3.

Chapter 4

1. (a) By the product rule there are $4 \times 4 \times 6 = 96$ terms in the complete expansion of $(a + b + c + d) \cdot (e + f + g + h) \cdot (u + v + w + x + y + z)$.

(b) The terms bvx and egu don't occur as summands in this expansion.

2. (a) $\binom{12}{9}$.

(b) $\binom{12}{9}(2^3)$.

(c) Let $a = 2x$ and $b = -3y$. By the binomial theorem the coefficient of a^9b^3 in the expansion of $(a + b)^{12}$ is $\binom{12}{9}$. So $\binom{12}{9}a^9b^3 = \binom{12}{9}(2x)^9(-3y)^3 = \binom{12}{9}(2^9)(-3)^3x^9y^3$, and the coefficient of x^9y^3 is $\binom{12}{9}(2^9)(-3)^3$.

3. $\binom{n}{n_1}\binom{n-n_1}{n_2}\binom{n-n_1-n_2}{n_3}\cdots\binom{n-n_1-n_2-n_3-\cdots-n_{t-1}}{n_t}$

$= \left(\frac{n!}{n_1!(n-n_1)!}\right)\cdot\left(\frac{(n-n_1)!}{n_2!(n-n_1-n_2)!}\right)\cdot$
$\left(\frac{(n-n_1-n_2)!}{n_3!(n-n_1-n_2-n_3)!}\right)\cdots\left(\frac{n_t!}{n_t!0!}\right)$

$= \frac{n!}{n_1!\,n_2!\,n_3!\cdots n_t!}.$

4. (a) $\binom{4}{1,1,2} = 12$.

(b) $\binom{4}{0,1,1,2} = 12$.

(c) $\binom{4}{1,1,2}(2)(-1)(-1)^2 = -24$.

(d) $\binom{4}{1,1,2}(-2)(3)^2 = -216$.

(e) $\binom{8}{3,2,1,2}(2)^3(-1)^2(3)(-2)^2 = 161{,}280$.

5. (a) $\begin{pmatrix} 10 \\ 2,2,2,2,2 \end{pmatrix} = (10!)/(2!)^5 = 113,400.$

   (b) $\begin{pmatrix} 12 \\ 2,2,2,2,4 \end{pmatrix}(2)^2(-1)^2(3)^2(1)^2(-2)^4$

   $$= [(12!)/[(2!)^4(4!)]](2)^2(3)^2(2)^4$$
   $$= 718,502,400.$$

   (c) $\begin{pmatrix} 12 \\ 0,2,2,2,2,4 \end{pmatrix}(1)^2(-2)^2(1)^2(5)^2(3)^4$

   $$= [(12!)/[(0!)(2!)^4(4!)]](2)^2(5)^2(3)^4$$
   $$= 10,103,940,000.$$

6. In each of parts (a)–(e), replace the variables by 1 and evaluate the results.
   (a) 2^3.
   (b) 2^{10}.
   (c) 3^{10}.
   (d) 4^5.
   (e) 4^{10}.

7. (a) $\displaystyle\sum_{i=0}^{n} \frac{1}{i!(n-i)!} = \frac{1}{n!}\sum_{i=0}^{n}\frac{n!}{i!(n-i)!} = \frac{1}{n!}\sum_{i=0}^{n}\binom{n}{i} = \frac{2^n}{n!}.$

   (b) $\displaystyle\sum_{i=0}^{n} \frac{(-1)^i}{i!(n-i)!} = \frac{1}{n!}\sum_{i=0}^{n}\frac{(-1)^i n!}{i!(n-i)!} = \frac{1}{n!}\sum_{i=0}^{n}(-1)^i\binom{n}{i} = \frac{1}{n!}(0) = 0.$

8. $n\begin{pmatrix} m+n \\ m \end{pmatrix} = n\dfrac{(m+n)!}{m!n!} = \dfrac{(m+n)!}{m!(n-1)!}$

   $$= (m+1)\frac{(m+n)!}{(m+1)(m!)(n-1)!}$$
   $$= (m+1)\frac{(m+n)!}{(m+1)!(n-1)!}$$
   $$= (m+1)\begin{pmatrix} m+n \\ m+1 \end{pmatrix}.$$

9. The sum is the binomial expansion of $(1+2)^n = 3^n$.

10. (a) $1 = [(1+x)-x]^n$

$$= (1+x)^n - \binom{n}{1}x^1(1+x)^{n-1} +$$

$$\binom{n}{2}x^2(1+x)^{n-2} - \cdots + (-1)^n\binom{n}{n}x^n.$$

(b) $1 = [(2+x)-(x+1)]^n.$

(c) $2^n = [(2+x)-x]^n.$

11. $\displaystyle\sum_{i=0}^{50}\binom{50}{i}8^i = (1+8)^{50} = 9^{50} = [(\pm 3)^2]^{50} = (\pm 3)^{100}$, so $x = \pm 3$.

12. $\displaystyle\sum_{i=1}^{3}(a_i - a_{i-1}) = (a_1 - a_0) + (a_2 - a_1) + (a_3 - a_2) = a_3 - a_0.$

13. $\displaystyle\sum_{i=1}^{n}(a_i - a_{i-1}) = (a_1 - a_0) + (a_2 - a_1) + (a_3 - a_2) + \cdots + (a_{n-1} - a_{n-2}) +$

$(a_n - a_{n-1}) = a_n - a_0.$

14. (a)
```
for i := 1 to 5 do
    for j := i + 1 to 6 do
        print(i, j)
```

(b)
```
for i := 1 to 4 do
    for j := i + 1 to 5 do
        for k := j + 1 to 6 do
            print(i, j, k)
```

Chapter 5

1. Let x_i, $1 \le i \le 5$, denote the amounts given to the five children.

(a) The number of integer solutions of $x_1 + x_2 + x_3 + x_4 + x_5 = 10$,

$x_i \ge 0$, $1 \le i \le 5$, is $\dbinom{5+10-1}{10} = \dbinom{14}{10}$. Here $n = 5$, $r = 10$.

(b) Giving each child one orange results in the equation $x_1 + x_2 +$

$x_3 + x_4 + x_5 = 5$, $x_i \ge 0$, $1 \le i \le 5$. There are $\dbinom{5+5-1}{5} = \dbinom{9}{5}$ ways to

distribute the remaining five oranges.

(c) Let x_5 denote the amount for the oldest child. The number of integer solutions of $x_1 + x_2 + x_3 + x_4 + x_5 = 10$, $x_i \geq 0$, $1 \leq i \leq 4$, $x_5 \geq 2$ is the number of solutions to $y_1 + y_2 + y_3 + y_4 + y_5 = 8$, $y_i \geq 0$, $1 \leq i \leq 5$, which is $\binom{5+8-1}{8} = \binom{12}{8}$.

2. Let x_i, $1 \leq i \leq 5$, denote the number of gumdrops for the five children with x_1 the number for the youngest. $(x_1 = 1)$: $x_2 + x_3 + x_4 + x_5 = 14$. Here there are $\binom{4+14-1}{14} = \binom{17}{14}$ distributions. $(x_1 = 2)$: $x_2 + x_3 + x_4 + x_5 = 13$. Here there are $\binom{4+13-1}{13} = \binom{16}{13}$ distributions. By the sum rule the answer is $\binom{17}{14} + \binom{16}{13}$.

3. $\binom{4+20-1}{20} + \binom{23}{20}$.

4. (a) $\binom{31}{12}$.

(b) $\binom{31+12-1}{12} = \binom{42}{12}$.

(c) There are 31 ways to have 12 cones with the same flavor. So there are $\binom{42}{12} - 31$ ways to order the 12 cones and have at least two flavors.

5. (a) 2^5.

(b) For each of the n distinct objects there are two choices. If an object is not selected, then one of the n identical objects is used in the selection. This results in $2n$ possible selections of size n.

6. $\binom{12}{4,4,4}\binom{22}{12}$.

7. (a) $\dbinom{4+32-1}{32}=\dbinom{35}{32}$.

(b) $\dbinom{4+28-1}{28}=\dbinom{31}{28}$.

(c) $\dbinom{4+8-1}{8}=\dbinom{11}{8}$.

(d) 1.

(e) $x_1 + x_2 + x_3 + x_4 = 32$, $x_i \geq -2$, $1 \leq i \leq 4$. Let $y_i = x_i + 2$, $1 \leq i \leq 4$. The number of solutions to the given problem is then the same as the number of solutions to $y_1 + y_2 + y_3 + y_4 = 40$, $y_i \geq 0$, $1 \leq i \leq 4$. This is $\dbinom{4+40-1}{40}=\dbinom{43}{40}$.

(f) $\dbinom{4+28-1}{28}-\dbinom{4+3-1}{3}=\dbinom{31}{28}-\dbinom{6}{3}$, where the term $\dbinom{6}{3}$ accounts for the solutions where $x_4 \geq 26$.

8. For the chocolate donuts there are $\dbinom{3+5-1}{5}=\dbinom{7}{5}$ distributions. For the jelly donuts there are $\dbinom{3+4-1}{4}=\dbinom{6}{4}$ distributions. By the product rule there are $\dbinom{7}{5}\dbinom{6}{4}$ ways to distribute the donuts as specified.

9. $230,230=\dbinom{n+20-1}{20}=\dbinom{n+19}{20} \Rightarrow n=7$.

10. Here we want the number of integer solutions for $x_1 + x_2 + x_3 + x_4 + x_5 + x_6 = 100$, $x_i \geq 3$, $1 \leq i \leq 6$. (For $1 \leq i \leq 6$, x_i counts the number of times the face with i dots is rolled.) This is equal to the number of nonnegative integer solutions to $y_1 + y_2 + y_3 + y_4 + y_5 + y_6 = 82$, $y_i \geq 0$, $1 \leq i \leq 6$. So the answer is $\dbinom{6+82-1}{82}=\dbinom{87}{82}$.

11. (a) $\binom{10+5-1}{5} = \binom{14}{5}$.

(b) $\binom{7+5-1}{5} + 3\binom{7+4-1}{4} + 3\binom{7+3-1}{3} + \binom{7+2-1}{2}$

$= \binom{11}{5} + 3\binom{10}{4} + 3\binom{9}{3} + \binom{8}{2}$,

where the first summand accounts for the case where none of 1, 3, 7 appears, the second summand for when exactly one of 1, 3, 7 appears once, the third summand for the case of exactly two of these digits appearing once each, and the last summand for when all three appear.

12. (a) The number of solutions for $x_1 + x_2 + x_3 + x_4 + x_5 < 40$, $x_i \geq 0$, $1 \leq i \leq 5$ is the same as the number for $x_1 + x_2 + x_3 + x_4 + x_5 \leq 39$, $x_i \geq 0$, $1 \leq i \leq 5$, and this equals the number of solutions for $x_1 + x_2 + x_3 + x_4 + x_5 + x_6 = 39$, $x_i \geq 0$, $1 \leq i \leq 6$. There are $\binom{6+39-1}{39} = \binom{44}{39}$ such solutions.

(b) Let $y_i = x_i + 3$, $1 \leq i \leq 5$, and consider the inequality $y_1 + y_2 + y_3 + y_4 + y_5 \leq 54$, $y_i \geq 0$. There are [as in part (a)] $\binom{6+54-1}{54} = \binom{59}{54}$ solutions.

13. (a) $\binom{4+4-1}{4} = \binom{7}{4}$.

(b) $\binom{3+7-1}{7}$ (container 4 has one marble) +

$\binom{3+5-1}{5}$ (container 4 has three marbles) +

$\binom{3+3-1}{3}$ (container 4 has five marbles) +

$\binom{3+1-1}{1}$ (container 4 has seven marbles)

$= \sum_{i=0}^{3} \binom{9-2i}{7-2i}$.

14. (a) $\binom{8}{2,4,1,0,1}(3)^2(2)^4.$

(b) The terms in the expansion have the form $v^a w^b x^c y^d z^e$ where a, b, c, d, e are nonnegative integers that sum to 8. There are $\binom{5+8-1}{8} = \binom{12}{8}$ terms.

15. Consider one such distribution: six books on each of the four shelves, which can happen in 24! ways. And clearly there are also 24! ways to place the books for any other such distribution.

The number of distributions is the number of positive integer solutions to

$$x_1 + x_2 + x_3 + x_4 = 24.$$

This is the same as the number of nonnegative integer solutions for

$$y_1 + y_2 + y_3 + y_4 = 20.$$

Here $y_i + 1 = x_i$ for all $1 \le i \le 4$.

There are $\binom{4+20-1}{20} = \binom{23}{20}$ such distributions of the books, and so $\binom{23}{20}(24!)$ ways in which we can arrange the 24 books on the four shelves with at least one book on each shelf.

16. (a) $\binom{5+12-1}{12} = \binom{16}{12}.$

(b) $5^{12}.$

17. Here there are $r = 4$ nested for loops and $1 \le m \le k \le j \le i \le 20$. So we're making selections, with repetition, of size $r = 4$ from a collection of size $n = 20$. Hence the print statement executes $\binom{20+4-1}{4} = \binom{23}{4}$ times.

18. Here there are $r = 3$ nested for loops and $1 \leq i \leq j \leq k \leq 15$. So we're making selections, with repetition, of size $r = 3$ from a collection of size $n = 15$. Hence the statement `counter := counter + 1` executes $\binom{15+3-1}{3} = \binom{17}{3}$ times, and the final value of *counter* is $10 + \binom{17}{3} = 690$.

19. The `begin-end` snippet executes $\binom{10+3-1}{3} = \binom{12}{3} = 220$ times.

    After execution the final value of *sum* is $\sum_{i=1}^{220} i = (220)(221)/2 = 24{,}310$.

20. Put one object into each container. Then there are $m - n$ identical objects to place into n distinct containers. This yields $\binom{n+(m-n)-1}{m-n} = \binom{m-1}{m-n} = \binom{m-1}{n-1}$ distributions.

21. (a)
```
for i := 0 to 10 do
    for j := 0 to 10 - i do
        print (i, j, 10 - i - j)
```

    (b) For all $1 \leq i \leq 4$ let $y_i = x_i + 2 \geq 0$. Then the number of integer solutions to $x_1 + x_2 + x_3 + x_4 = 4$, where $x_i \geq -2$ for $1 \leq i \leq 4$, is the number of integer solutions to $y_1 + y_2 + y_3 + y_4 = 12$, where $y_i \geq 0$ for $1 \leq i \leq 4$. We use this result to write the following snippet.
```
for i := 0 to 12 do
    for j := 0 to 12 - i do
        for k := 0 to 12 - i - j do
            print (i, j, k, 12 - i - j - k)
```

22. If the summands must all be even, then consider one such composition:
$$20 = 10 + 4 + 2 + 4 = 2(5 + 2 + 1 + 2).$$

    Note that $5 + 2 + 1 + 2$ is a composition of 10. Also, each composition of 10, when multiplied through by 2, is a composition of 20, where each summand is even. So the number of compositions of 20 where each summand is even equals the number of compositions of 10—namely, $2^{10-1} = 2^9$.

23. Each such composition can be factored as k times a composition of m. So there are 2^{m-1} compositions of n, where $n = mk$ and each summand in a composition is a multiple of k.

Index

www.ingramcontent.com/pod-product-compliance
Lightning Source LLC
Chambersburg PA
CBHW070502220526
45467CB00002B/533